PASSION
of the
GOLD PYRAMID

by

Yolanda Fierro

For permission requests, write to the publisher, addressed "Attention: Permissions Coordinator" carol@markvictorhansenlibrary.com

Quantity sales special discounts are available on quantity purchases by corporations, associations, and others. For details, contact the publisher at carol@markvictorhansenlibrary.com

Orders by U.S. trade bookstores and wholesalers. Email: carol@markvictorhansenlibrary.com

Any references to historical events, real people, or real places are used fictitiously. Other names, characters, places, and events are products of the author's imagination, and any resemblance to actual events or places or persons, living or dead, is entirely coincidental.

Creative contribution by Lyn South
Cover Design & Book Layout - DBree, StoneBear Design

Manufactured and printed in the United States of America distributed globally by markvictorhansenlibrary.com
New York | Los Angeles | London | Sydney

MVHI

ISBN: 979-8-88581-145-3 Hardback
ISBN: 979-8-88581-146-0 Paperback
ISBN: 979-8-88581-147-7 eBook
Library of Congress Control Number: 2024903246

I wrote this book on behalf of
James A. Onan
to honor his father
James Paul Onan

9/25/1937 to 7/16/2023

Foreword
James A. Onan

Everyone has a hero growing up and mine is a man who is larger than life itself, my father the builder of the Gold Pyramid House.

The impact he had on me and anyone who had the pleasure of meeting him was life changing. His visions were sometimes crazy to me as a young boy but as any good son should do, I respected him and followed his direction.

As children we don't always see eye-to-eye with our parents and it's not until we get older do we start to understand why they do the things they do. It's not until our eyes mature do we see their vision.

I never knew that one day I would share in his vision for the Pyramid and it would be one of my biggest challenges and achievements of my life.

I would come to believe in the mysteries surrounding the power of the pyramid and the healing affects it has on the body. That's why I took my father's vision even further and re-opened the house for visitors to be able to experience a little taste of Egypt and feel firsthand the energy and vibrations of the pyramid.

I came to realize that this three-sided house is more than a just a home but it's a testament that you can achieve anything if you put your mind to it. Once you find your passion believe in yourself enough, then anything is possible.

It's all in what you let your mind believe.

Yolanda Fierro

Table of Contents

Yolanda Fierro

Chapter One

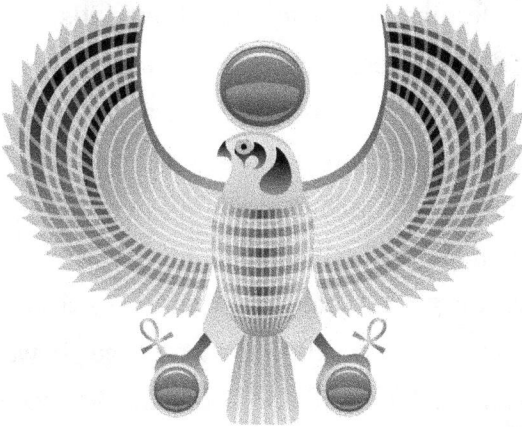

n Lake County, Illinois, there is a small village called Wadsworth, a picture-perfect Midwestern town where young families liked to raise their kids, and retirees spent their days puttering around in their gardens or fishing on the nearby lake. There were the typical amenities—grocery stores, doctors' offices, and schools.

And there was a golden Egyptian pyramid—an honest-to-goodness five story tall, seventeen thousand square foot pyramid with a 50-foot tall 200 tons of steel and concrete statue of Ramses II guarding the entrance. It was August, and the heat baked them like someone had shoved them into an oven and closed the door.

A small, rural town outside of Chicago was a strange place to find a pyramid, but there it stood in all its glory. Towering over Anthony Johnson as he scowled at the structure from the gravel parking lot and wished he had agreed to help his friend Peter clean out forty years of hoarded clutter at his parents' place. It was a total Tom Sawyer move on Pete's part—getting people to help with the most arduous and disgusting tasks he didn't want to do himself and making it seem like the helpers were getting the better end of the deal.

Right now though, Anthony would much rather be sifting through piles of old newspapers, broken tools, and mystery pieces from old building projects than touring some lame tourist trap. But his parents had insisted, and so here he was, trudging toward the front door of a pyramid that shouldn't be in the middle of Nowheresville, Illinois.

"Isn't this amazing?" his mother gushed as she took in the sight of the towering structure. "I've always wanted to see Egypt."

"It's something all right," Anthony muttered under his breath, not wanting to admit that, despite his grumbling, he was actually kind of impressed by the size and scope of the pyramid. He rubbed his temples. He just got over a two-day migraine and his head still hurt. "And to be fair, you're around the corner from cornfields so you're still not seeing Egypt."

"Anthony, we drove all the way out here," Mom said with a heavy sigh. "The least you can do is go inside and take a look around. Besides, your sister is so excited to see this place, and I don't want you ruining this outing for her."

Anthony's dad gave him a sympathetic pat on the shoulder. "Come on, son. Take one for the team. It'll be fun. I promise."

"Anthony, come on!" his sister Jenna called out as she

ran towards the front door of the pyramid. "I want to see what it's like inside!"

With a shrug of his shoulders, Anthony followed Jenna down the path to the pyramid. He was about to enter the front door when he was stopped by a smiling man in a khaki safari outfit. Anthony suppressed an eye roll, already feeling like he was in some kind of cheesy, 80s action-adventure movie.

"Welcome to the Gold Pyramid," the man said with a smile. "I'm Harold, and I'll be guiding your tour through our unique and amazing structure."

"Hi, Harold," Anthony's mom said as she handed him their tickets. "We're so excited to be here."

"I can see that," Harold said with a chuckle. He leaned down to address ten-year-old Jenna at eye level. "Are you ready for an adventure, young lady?"

"You bet!" Jenna said with a wide smile. "I've never been in a pyramid before."

"Then you're in for a real treat," Harold said as he ushered them through the front door. "This way to the treasure room."

"It's not every day you get to see the largest gold structure in North America," his mother said.

Harold stood just outside the entrance and pointed at the outside of the pyramid. Anthony just wanted to get inside where, hopefully, the air conditioning was on full blast, and they might get some relief from the heat.

Their tour guide, however, seemed oblivious to the heat

and didn't miss a beat as he chatted away. "From the outside, The Gold Pyramid looked like it was plucked from the sands of Egypt and dropped in the middle of Illinois. You may think this is just a museum, but it's not. It's an actual house that the owners live in. We are obviously open for tours and we're so excited you could be here with us today."

Anthony followed his family and Harold through the front door, barely paying attention to the man's running commentary. He was too busy gawking at his surroundings. He didn't expect the inside of the pyramid to be even more incredible than the outside. The place conjured an otherworldly feel, and Anthony wasn't even mildly interested, and he was annoyed with himself that he hadn't brought his ear buds so he could listen to music while pretending to pay attention.

The first room inside the entrance was a foyer filled with replica of Egyptian artifacts: a large turquoise colored urn stood amid small and large statues. The floor was a light tile edged in gold trim, and Anthony could see more replicas inside the building through the numerous windows in the foyer. Hieroglyphics of all kinds were painted above the windows and across the walls.

There was also an old Treasures of Tutankhamun poster from Chicago's Field Museum of Natural History. Harold even had something to say about the history-making tour of Egyptian artifacts. "Back in the mid to late seventies, The

Treasures of Tutankhamun toured the United States, and it was a huge deal. People stood in line for hours—sometimes even overnight. In August of 1977, Chicago's Field Museum of Natural History hosted the tour, and more than one million people came to see it during its four months stay."

"Wow," Jenna said, her eyes wide with wonder. "I wish I could've seen it."

"Oh, it was something all right," Harold said. He turned and motioned for them to follow him.

The only thing Anthony remembered about King Tut wasn't from school, it was an old Saturday Night Live musical number he saw online with comedian Steve Martin singing his song about the Boy King.

Jenna ran over to inspect the large turquoise colored urn and Mom reminded her not to touch anything before putting her hand on it.

"When was it built?" Dad asked as he inspected a large sarcophagus. It looked like a replica of King Tut's with its gold and black head and mask and the colorful robes of red, gold, and orange.

A small table of an Egyptian servant, clad in a gold loincloth and holding a tray with two bottles knelt beside the sarcophagus. Anthony leaned over to inspect the bottles. One was a small, pyramid-shaped bottle filled with mineral water, and the other was a tall bottle of vodka.

Anthony raised an eyebrow and Harold quickly

explained. "We have a natural mineral spring that erupted from the ground in the middle of the pyramid while digging for an inground pool. The water was found to have calcium, magnesium, and potassium that combined to make the water a natural electrolyte."

"To answer the question about the construction," Harold turned back to Dad. "Building on The Gold Pyramid began in 1977 and was completed in 1982," Harold said proudly. "It's made entirely of reinforced concrete, and the exterior was, at one time, covered in over 24 karat gold leaf. It was believed to be the largest 24 karat gold-plated object in North America. They had numerous complaints about how the sun's rays, at certain times of the day, were blinding to drivers on a nearby highway and pilots flying to and from Chicago O'Hare and Milwaukee airports. Mr. Omen was asked to remove the gold."

"Oh, no. What did he do?" Jenna asked.

"He refused. Instead, he placed a huge dome over the entire house. That stopped all the glare from the sun. Several years later a tornado ravaged the neighborhood and knocked one of the steel beams into the house causing structural damage. The gold had to be removed to repair the structure and was melted down and turned into coins and a gold like coating was applied to exterior of the house to give it that gold look.

"Whoa," Dad said, looking startled. "That's a lot of gold

coating. It makes sense, though, that the sun would reflect off the structure like that."

"The owner, Paul Omen, Sr., and his wife had a dream to build this pyramid and they spent over one million dollars achieving that dream." Harold continued.

"It's pretty crazy to build a pyramid out in the country." Jenna said, her eyes wide as she looked around.

"Yes, it is pretty crazy." Harold agreed with a chuckle. "But that's what makes this place so special. You never know what you're going to find when you visit The Gold Pyramid."

Just past the foyer was a sort of reception room but it was unlike anything Anthony had ever seen. In the middle of the carpet, inside a red-velvet rope barrier, lay a gold-framed piece of glass over what looked like a large hole in the floor. Above the glass were four long rods joined at the top to form a pyramid. On either side of the hold lay two, small yet ornate sphinxes and standing at the top of the glass was a golden statue of an Egyptian woman with her arms outstretched as though she were offering a blessing or welcoming visitors.

"That's Isis," Harold said, noticing Anthony's fascination with the statue. "Isis was the ancient Egyptian goddess of healing and protection."

"And that hole in the floor?" Jenna asked, her curiosity peaked as she leaned over the glass to get a better look.

"Remember the bottle of mineral water you saw out there in the foyer? That's a glimpse into the mineral water spring that flows under the pyramid," Harold explained. "The glass allows our visitors to see the water as it flows beneath us. How we found the mineral spring is quite a story that I want to save for after the rest of the tour." He glanced at Jenna. "I bet you're excited to see the rest of the treasures inside. Then, we have a replica of King Tut's tomb to explore."

"Wow! This is going to be so cool." Jenna nodded, and they followed Harold as he led the way. The next room was larger and filled with all sorts of ancient Egyptian artifacts.

There were statues of Pharaohs, jewelry, coins, and furniture. Anthony thought it all looked really cool like the replicas belonged in a real Egyptian tomb.

There were several chairs in the room, replicas of ancient Egyptian chairs. They were made of wood and adorned with intricate carvings. The seats were upholstered in a crimson fabric and the legs were curved and tapered. Each chair had a high backrest that was decorated with gold leaf. The arms of the chairs were also carved and ended in paw-like feet. These chairs would have been fit for a king or queen, and Anthony couldn't help but wonder if Ramses the Second might have sat in chairs just like them in ancient times.

The replicas of the ancient Egyptian coins and jewelry were impressive. Dad—an avid coin collector—fussed over the coins. He commented on the fine details of the engraving and the beautiful workmanship. The replicas of the jewelry were equally as stunning, and Mom was drawn to a necklace that featured a large, blue stone set in a scarab. She commented on the exquisite craftsmanship as she watched the light glitter in the stone.

After exploring the Egyptian artifact replica displays in the pyramid house, Harold led them outside to a smaller concrete structure that resembled a bunker. A metal staircase led into the structure which turned out to be an exact replica in size and scope of King Tut's tomb. It featured hieroglyphs painted on the walls.

The best thing about it, Anthony thought, was it was much cooler than he expected which helped to soothe his aching head.

There were replicas of the treasures found in the young pharaoh's tomb—a senet game, statues of gods and goddesses, and a replica of King Tut's iconic death mask. What did it look like?

In the center of the tomb, encased in glass, was an enormous replica of the pharaoh's open sarcophagus. "Look!" Jenna squealed as she gazed inside. "There's a mummy!"

"It's not real," Harold said with a chuckle. "But we thought it might be fun to give visitors a glimpse of what it might have been like to explore Tutankhamun's tomb for the first time, mummy included."

Jenna was fascinated by the tomb and asked Harold a million questions. He answered them all with patience and good humor. It was clear that he loved his job and that he was passionate about sharing history with others. Anthony had to admit, the detail and accuracy of the replica tomb—and all of the items in the pyramid house—was impressive.

"What was King Tut like? Was he a good pharaoh?" Jenna asked as she gazed at the replica of Tutankhamun's death mask.

"There's still much that we don't know about him," Harold replied. "But from what we do know, he seems to have been a fairly average young man who became pharaoh at a very young age. He was only nine years old when he took the throne."

"What happened to him? How did he die?" Jenna wanted to know.

"That's still a bit of a mystery," Harold said. "Most historians believe that he died from an infection or disease, but we're not really sure. Some people think that he might have been murdered."

Jenna was quiet for a moment, taking in all of the information. Then she said, "In school, we learned a little bit about how the pharaohs lived in ancient times. My teacher said they believed in all kinds of gods and goddesses."

"Yes, that's correct," Harold said. "The Egyptians had a complex religion with many different gods and goddesses. They believed that these deities controlled everything in the world, from the weather to the Nile River. They also believed that the pharaohs were gods themselves."

"That's right," Dad said. "I remember learning about that when I was a boy in school, too. The Egyptians believed that the pharaohs were reincarnated gods who ruled over them."

Jenna looked at the hieroglyph paintings on the walls of the tomb. "Are these paintings real? I mean, were these pictures the same ones that were painted on the walls of King Tut's tomb?"

"Yes," Harold replied, enthusiastically. He walked over to a wall and pointed at the first wall. "Each of the walls in King Tut's tomb were painted to depict a specific part of the pharaoh's journey into the afterlife. And Mr. Omen spared no expense or detail in getting the paintings in our version of King Tut's tomb exactly right." He pointed at the first wall. "This wall depicts the funeral procession as it makes its way to the tomb."

"What about this wall?" Anthony asked, finding himself getting caught up in the history lesson despite himself. He pointed to a wall that depicted what looked like a battle scene.

"That's the pharaoh's arrival in the underworld," Harold explained. "He has to overcome all sorts of obstacles and challenges in order to reach the afterlife. And the last one over there depicts King Tut's arrival in the Afterlife."

"Weren't there a lot of things put in the tomb with King Tut? Like food and furniture and stuff?" Jenna asked, her eyes wide.

"Yes, there were," Harold replied. "The ancient Egyptians

believed that the pharaohs would need all of those things in the afterlife, so they put them in the tomb. In fact, Howard Carter, one of the men who discovered the tomb, spent over a decade clearing items out of the tomb and cataloging them."

Anthony realized he wasn't as bored as he thought he was going to be. Still, he was ready to leave and go explore the rest of the pyramid house. He was about to say as much to Jenna and Dad when Harold said something that made him pause.

"Of course, not everything that's important about the pyramids can be displayed like a mummy or some paintings," Harold said, his voice lowered as if he were sharing a secret. "There are some things that are too precious, too valuable to put on display."

"What do you mean?" Jenna asked, her eyes wide.

"I mean that the pyramids are worth more than just for historical significance," Harold replied. He leaned in close and said, "There are secrets about the pyramids that have yet to be fully appreciated."

"What kind of secrets?" Dad asked, his interest piqued.

"The kind of secrets that could change the world," a man's voice replied, his voice full of intrigue. They all turned to see an older man standing at the entrance to King Tut's chamber. He was tall and thin, with a kind smile and sparkling eyes. "Are you ready to hear it?"

Yolanda Fierro

Chapter Two

"Ah. We're in for a special treat today, folks," Harold said. "This is Mr. Paul Omen, Senior. He built the pyramid house and this replica tomb."

Harold introduced each member of the Johnson family, and Paul Omen smiled widely as he shook their hands. "It's a pleasure to meet you. We're so happy you joined us here at the pyramid today."

"Likewise," Dad replied. "This is quite a place you have here. Harold has done a great job of showing us around and explaining everything."

"Thank you," Harold said. "I enjoy my job here very much. It's a labor of love."

"And it shows," Mom added. "This place is amazing. We're really enjoying our visit."

"Glad to hear it," Paul said. "The power of this place is what is truly amazing."

Mr. Omen's comment piqued Anthony's curiosity. As far as he knew, the pyramid house and King Tut's replica tomb were interesting museum but nothing more. "What kind of power?"

"The power to change lives," Paul replied. "The power to give people hope."

"I'm not sure I understand," Anthony said, glancing at Harold. Surely, he would set this man straight and explain that the pyramid house was nothing more than a replica of an ancient Egyptian tomb—an impressive one, to be sure, but nothing magical or powerful.

But Mr. Omen simply nodded and gave Anthony a knowing smile. "Ah. A skeptic. I always love a good challenge." He turned to Harold. "Shall I show them?"

"I think that's a splendid idea," Harold replied with a big smile.

Harold led the way back into the pyramid house. Mr. Omen chatted amiably with Anthony's parents as they walked. When they reached the main room, he gestured for them to sit.

"As Harold probably told you, I built this pyramid house as a replica of an The Great Pyramid of Giza," he began. "But it is more than that. It is a place of power."

"Yeah, you said that before but what kind of power?" Anthony asked, not bothering to hide the skepticism in his voice.

"Pyramids have been used for centuries as a source of energy and healing," Paul explained. "The power of the pyramid can be harnessed to help people in many different ways. Back in the 1970s, people were looking for all kinds of alternative forms of religion and spirituality. One of the beliefs that gained popularity during this time was the idea that pyramids have power or energy that can be harnessed."

"Harnessed? You mean like electricity?" Jenna asked, her eyes wide.

"In a way, yes," Paul replied. "But it's not exactly the same kind of power. The power I'm speaking of is more like a positive force or energy that can be used to help people in their lives."

"How does it work?" Mom asked.

"There are many ways to harness the power of the pyramid," Paul explained. "Some people believe that meditating in a pyramid can help to promote healing and well-being. Others use pyramids to help manifest their desires or goals. I've even heard of people using pyramids to help them connect with their spiritual guides or other beings."

"So, you're saying that this pyramid house has some kind of magical power?" Anthony asked, still not convinced.

"Not exactly," Paul replied. "I prefer to think of it as a place of positive energy. A place where people can come to make changes in their lives."

"And you really believe that?" Anthony asked.

"I do," Paul replied. "I've seen the power of this pyramid firsthand. I've seen it change people's lives."

"How?" Anthony narrowed his eyes at the older man, still not sure if he believed him or not.

"I've seen people come into this pyramid who were struggling with all kinds of issues," Paul explained. "I've known people whose migraine headaches were completely cured after spending time in the pyramid. I've had people tell me that their anxiety and depression disappeared after spending time here. Peoples' eyesight has improved, too."

"I'm not sure I believe that," Anthony said. "It sounds like a bunch of hocus-pocus to me."

"Does it now? Well, I suppose you'll just have to experience it for yourself then," Paul replied. He turned to Anthony's parents. "Would either of you mind if I gave young Anthony, here, a special tour? I have some things to show our young skeptic."

Dad and Mom agreed and, with Jenna in tow, followed Harold outside to grab a bite to eat at the concession stand. Anthony followed Mr. Omen back into the pyramid house.

"I thought I'd show you something a little more, shall we say, hands-on?" Paul said with a twinkle in his eye.

Anthony frowned, not sure what the older man meant. He followed him into one of the rooms off of the main chamber which turned out to be an office. Mr. Omen sat down behind a desk and motioned for Anthony to sit in one of the chairs in front of it.

"So, you're a skeptic when it comes to metaphysical things, huh?" Paul asked.

"I guess you could say that," Anthony replied. "I've never really believed in that stuff that can't be explained by science."

"Good! I like a healthy dose of skepticism. It keeps us from getting taken advantage of by people who would exploit our beliefs or our desire to believe in something," Paul said. "But I think you'll find that there's more to this world than what traditional science can sometimes explain. And there are things that can be tested—powers of the pyramid, for example—that traditional science doesn't know how to measure yet. And even if they did know how to measure it, there are some who openly reject anything having to do with the powers pyramids are reported to have.

"I guess I can see that," Anthony replied. "But it's still hard for me to wrap my mind around the idea of metaphysical powers and pyramids being more than just interesting structures."

"I understand," Mr. Omen said. "It takes an open mind to really see and believe in the power of the pyramid. But I noticed you kept rubbing your forehead as you toured the house today. And you kept rubbing it while you were in the tomb."

"Yeah." Anthony rubbed his forehead, feeling a dull ache forming again. "I get headaches a lot."

"Migraines?" Paul asked.

"Sometimes," Anthony replied.

"Do you get them often?"

"I do. I've seen doctors and they prescribed meds, which help, but the headaches keep coming back."

Mr. Omen leaned forward and rested his forearms on the desk. He peered intently at Anthony and said, "I think I might have a solution for you."

Anthony's eyebrows shot up in surprise. He hadn't expected the older man to offer him any help, let alone a solution for his headaches.

"What do you mean?" he asked cautiously.

"The power of the pyramid can help with all kinds of ailments, including migraines," Paul explained. "I've seen it happen time and time again. I think if you spend some time in the pyramid, you'll find that your headaches will start to go away."

Anthony was dubious, but he was also desperate for anything that might help with his chronic headaches. He'd been dealing with them for years and nothing had ever worked for more than a few weeks, if that.

"I don't know," he said uncertainly. "It's just so hard to believe."

"I understand," Mr. Omen replied. "But I think it's worth a try. Would you like to know more about how the pyramids can heal people?"

Anthony found himself curious, despite his skepticism. When he hesitated, Mr. Omen smirked playfully at him and said, "Come on. What do you have to lose except a little bit of time and your headaches?"

Anthony shrugged. "I guess you're right. I don't have anything to lose."

"Great! Let me show you some of my research and the studies that have been conducted on pyramids and their power."

He pulled several thick manila folders out of a nearby file cabinet and plopped them down on the desk. The folders were frayed around the edges and stained in various places with what looked like coffee mug rings. They looked like they had been used for decades. He handed one folder to Anthony.

"Start with these."

Anthony opened the folder and found it filled with everything from clippings from magazines to printed academic and scientific papers. He leafed through them, his eyebrows gradually creeping up his forehead in response to what he was reading.

He learned that in the 1970s, a new movement began to gain popularity in the United States—the power of the pyramids. The idea was that pyramids had special powers that could be used for healing, protection, and even prosperity. People built small pyramid structures in their homes and yards, and some even slept inside them, hoping to be able to tap into the power of the pyramid. There was something about the pyramids' shape that generated a powerful energy field.

Researchers created scale models of The Great Pyramid of Khufu and the Pyramids of Giza and began experimenting. By aligning the pyramids to the magnetic poles, they began studying the pyramids effects on everything from meditation to plant growth.

The results seemed mixed, but there was enough evidence to suggest that the pyramids did have some sort of power. The most promising results came from studies on the healing properties of the pyramid. It appeared that the pyramid could help with a variety of ailments, including migraines, arthritis, and even cancer.

Anthony grew more and more intrigued. He had never heard of any of this before. He had always thought of pyramids as ancient monuments, not as objects with special powers. But the more he read, the more he realized that there might be something to what Mr. Omen was saying.

He looked up to find the older man watching him intently. He had the second manila folder open in front of him, and the slight, patient smile never left his face.

"Have a look at this," Mr. Omen said, handing him several pages from the folder. "I think you'll find it interesting."

Anthony took the papers and began to read. This one was filled with testimonies from people who used pyramid power for healing. There were stories of migraines vanishing, arthritis pain disappearing, and even cancer tumors shrinking.

"Are these for real? I mean, actual documented medical cases?"

"They are indeed. These are real people with real problems who found relief in the pyramid."

Anthony was silent for a moment, his mind racing. Could it be true? Could the pyramid actually have healing powers? He thought back to his own migraines and the pain he had been in for years. What if there was a way to get rid of them for good?

"Hey, Anthony?" Mr. Omen's voice interrupted his thoughts. "How's your head?"

Anthony stopped reading and looked up at Mr. Omen. He was about to say that his head still hurt but, after a moment's consideration, he realized that it didn't. He didn't feel any pain or aches. In fact, he had almost completely forgotten about his migraines since he had sat down in Mr. Omen's office in his massive pyramid house.

"It's . . . it's fine," he said slowly, still trying to wrap his mind around what was happening. "My head doesn't hurt."

Mr. Omen nodded and smiled even wider. "Huh. Doesn't hurt at all?"

"No. I had a massive migraine last week that was just starting to get better. Until today. Then, my head started hurting again this morning. And it kept hurting right through the tour of the tomb but since we came in here, it's been fine."

"Interesting. Maybe it's the pyramid."

"Maybe," Anthony said, still not sure what to believe. "And maybe it's the fact that I'm not standing out in the blazing sun anymore. Or maybe my meds are working okay today."

"Or maybe it's the pyramid," Mr. Omen added with a laugh. Then, he gave Anthony a sincere look. "I'm glad you're feeling better. Let's get back to the documents. There's so much more for you to read before we go any further."

Anthony nodded and turned his attention back to the papers in his hand. He had a feeling he was going to be reading for a while. As he continued to read, he realized that there was a lot more to the pyramid than he had first thought. He held up a document for Mr. Omen to see. "This one's talking about some plants you started growing here at the pyramid that really took off."

"That's a very interesting case." Mr. Omen peered at the page before leaning back in his seat and gazing at Anthony. "We had a local college conduct a study on the power of the pyramid on plants and guess what they found?"

Anthony flipped through the pages until he found the reference to the results of the study Mr. Omen was talking about. "It says here that they were astonished to find that the plants grown here at your pyramid house grew three times faster than the plants grown at the university. And they were healthier too."

"Exactly. The university couldn't figure it out. They thought maybe we were using some sort of chemical or fertilizer that they didn't know about. But we weren't. We were just using regular potting soil and water." He laughed. "Well, actually we were giving it very special water."

Anthony shook his head, still trying to wrap his mind around what he was hearing. Could it be true? Was the pyramid really that powerful? "Wait. The pyramid is magic, but you have magic water, too?"

"That's right, my boy." Mr. Omen grinned at him like a schoolboy with a secret. "The water here at the pyramid is special because it's been charged with pyramid power."

"Charged with power? What do you mean?"

"Well, if you'll come with me, I'll show you."

Chapter Three

J aul led Anthony back to the foyer to where the hole in the floor looked down into the mineral spring that flowed beneath the pyramid house.

"We didn't exactly find a spring." Mr. Omen gave Anthony a wry smile. "It's more like the spring found us after we built the pyramid."

Anthony frowned, not understanding what the man meant. "I don't get it. What do you mean the spring found you?"

"As we built the pyramid house, the spring bubbled up from the center of the pyramid and began flowing," Paul explained. "It was like the pyramid was somehow drawing the water to us. So, we decided to use the geothermal spring to heat the house. We built a system of pipes that carries the water throughout the pyramid, and fans circulate the warm air."

"So, you're saying the pyramid house is heated by a natural spring?" Anthony asked, his eyebrows raised in disbelief.

"Yes," Paul replied. "And that's not all. We also bottle

the water from the spring and sell it as Gold Pyramid natural mineral water. It's said to have healing properties."

"Let me get this straight," Anthony said, still not sure he believed what Paul was telling him. "You're saying that the pyramid house is heated by a natural spring and the water from that spring has healing properties?"

"Yes. Of course. I've mentioned how the pyramids of Egypt are said to have a number of benefits, including improving health, sleep, and meditation. It's also said the power of the pyramids help heal cuts and burns and heal headaches. After the spring water began bubbling up, we had it tested. We were ecstatic to learn it is very high in calcium, magnesium, and other minerals that make it a natural electrolyte. So, the minerals in our water—brought to us by the power of our very own pyramid—are proven to play a major role in preventing heart disease and age-related bone loss."

Anthony frowned. "I get how those minerals in the water help people's health. I mean, that's just science. We already know important minerals like calcium and magnesium are good for us. But I don't see how the pyramid house itself has any power."

Mr. Omen nodded. "A true pyramid produces polarized signals that, in turn, create a form of electrical power. This power is said to have unique effects on the surrounding environment and that's why pyramids can even provide amazing health benefits to people. I know you're skeptical,

Anthony, but I think if you learn about the powers of this pyramid house, you might find that it can change your life."

As Anthony stared down at the aquifer that had mysteriously appeared beneath Paul Omen's Gold Pyramid, he had more questions than answers. How could it be possible that such an incredible natural phenomenon was occurring right here in this small midwestern town?

Anthony had always been skeptical about claims of supernatural power or healing ability, and he found it hard to believe that the aquifer could have any real effect on anyone's health or well-being. Still, he was beginning to get the feeling that there was something undeniably special about this place, and his curiosity bubbled up inside of him as he gazed down into its sparkling waters below.

Paul Omen gazed down at the aquifer with a sense of wonder shining in his eyes. From what Anthony could tell, the older man had no doubt that this natural phenomenon was nothing short of miraculous, and he launched into an explanation of just how it had appeared beneath his golden pyramid.

"An aquifer is an underground reservoir of water that can be found in porous rock formations or layers of sediment. It is typically formed through the natural filtering and storage of rainwater, groundwater, or melted snow. In some cases, aquifers are believed to have special healing properties due to the presence of certain minerals in their waters. Did you know that?"

Anthony grinned sheepishly. "Sorry. The last time I had geology in school, I think I was twelve."

"Well," he continued, "this aquifer is unique because—as I mentioned before—it sprang up right beneath my pyramid house after it was built. Did you know the house was originally built on an island?" He waved out the windows. "All of this was just an old gravel pit and, originally, there was only a few feet of water toward the back of the property."

"You built this all in a gravel pit, huh?"

"Sure did. The original plan was to spread clay over the bottom of the gravel pit, then fill it with well water. But when we finished the pyramid house construction, it was the craziest thing. It was as though the aquifer was drawn to the pyramid."

Anthony frowned. "That does seem weird. Were there any natural springs in the area already? I mean, maybe you guys just got lucky and hit an undiscovered water source."

"We looked at the water table maps, and they didn't list any natural aquifers in the area. So, it makes sense that the powerful energy and vibrations created by the pyramid structure is what drew the water to it. The aquifer is a clear, tangible sign of its powers, and I believe that it has the ability to not only heal and restore people, but to transform them as well."

As Anthony listened in awe to Paul's explanation, he couldn't help but feel deeply touched by the incredible story unfolding before him. But he was a skeptic at heart. His granddad often said to him, "Son, if something seems too good to be true, it probably is."

The more he heard about the pyramid house, and the underground aquifer, the more he was determined to find a logical explanation for what he was seeing here. Still, as he looked down at the aquifer he wondered if there really was a powerful energy source that pulled the water to the house.

And, if there was, Anthony couldn't help but wonder if this place might just hold the key to unlocking something truly remarkable within himself, too. Skeptical but intrigued, Anthony continued to listen to Mr. Omen's story. "We have welcomed visitors to our Gold Pyramid house for almost forty years. Every single one of them have been intrigued by this natural mineral water spring that, somehow, found its way to our house. We've had the water tested. It's high in calcium, magnesium, and lots of other minerals. It's a natural electrolyte, is what it is."

"Electrolyte? You mean like a sports drink?" Anthony asked.

"Sort of." Mr. Omen grinned. "It means this water is good for you, son," he said with a wink. "But don't take my word for it. Just try it yourself."

Mr. Omen gestured for Anthony to wait as he went to the kitchen and returned with a triangular shaped bottle with a blue label. On the label, was a picture of the Pyramid House, and the name Gold Pyramid Natural Mineral Water.

Mr. Omen offered the bottle to Anthony with a smile. "Go on. Try it."

Anthony opened the bottle and took a tentative sip of the mineral water. He was pleasantly surprised by its fresh taste as he swirled it around in his mouth before swallowing.

"Well?" said Mr. Omen expectantly.

Anthony grinned. "It's good. I like it."

"It should be." He laughed. "I sell the water by the bottle and at wholesale to local stores all around Illinois, so I make sure it's good!"

"It is really good mineral water. Light. Crisp. But does it really have special powers or healing abilities?" Anthony asked, curious.

Mr. Omen nodded solemnly. "It's not just my belief that it does. Scientific studies have shown the benefit of natural mineral water and electrolytes on people's health. In fact, studies have shown that the minerals in the water can reduce inflammation, improve sleep quality and even combat obesity."

Anthony was intrigued. He had never heard of anything like that before. "Really? Wow."

"Of course. One study from a reputable scientific journal showed that people who drank mineral water regularly for six weeks had improvements in multiple health markers, including blood pressure and inflammation. In fact, the studies show that mineral water lowers a person's LDL— the bad kind of cholesterol— and increases the HDL levels. That's the good kind of cholesterol. It also helps keep your heart healthy."

"Ah, wonderful. Thank you, Mary." Mr. Omen turned to Anthony. "David Hanson owns a local sports bar called Wide World Sports. He was one of the first people to start carrying our mineral water. I'd like for you to meet him and hear what he has to say about the quality of the water and how much his patrons love it."

Anthony nodded intrigued. "Sure," he said, curious to learn more about this natural mineral water and its healing properties.

Mr. Omen led Anthony through the main entrance of the Gold Pyramid House and towards another door that led out into the open air. As they stepped out onto the large deck overlooking the replica of King Tut's tomb at the front end of the property, Anthony saw a tall man with salt and pepper hair, wearing a pressed button-down shirt and khaki pants.

"David, so good to see you," Mr. Omen said, grinning as he extended his hand in greeting.

"Good to see you, too, buddy." Mr. Hanson returned the smile and enthusiastically grasped Mr. Omen's outstretched hand in an obvious display of camaraderie. "My business manager at the sports bar said we needed to place another order of Gold Pyramid Water. I told her that I needed to stop by and see my old friend, Paul, anyway, so I thought I'd place the order in person."

"I'm so glad you did," said Mr. Omen. "I wanted Anthony to meet you and hear about how much your patrons love our mineral water." He gestured to Anthony and smiled warmly. "This is my new friend, Anthony. He is visiting our Gold Pyramid House with his family, and I think he got a lot more than he bargained for today. I've been talking his ear off about the house and the healing properties of our mineral water."

"I'll bet you have. It's not easy to stop the Omen Pyramid discussion train once it gets goin'." Mr. Hanson laughed, extending his hand again to Anthony. "I'm David Hanson, owner of Wide World Sports over on the north side of town. Glad to meet you, Anthony."

Anthony shook his hand and smiled politely. "Nice to meet you too, Mr. Hanson."

"I see you've had a chance to sample the wares." He nodded at the Gold Pyramid Water bottle in Anthony's hand. "What do you think of it?"

"To be honest, I'm kind of surprised," Anthony replied. "It's really good. Some mineral water can taste a little strange or have a weird smell, but this is really smooth."

Mr. Hanson nodded approvingly. "My customers always say that they love it." He paused, looking thoughtful for a moment before continuing. "And, you know what? I have a lot of patrons who ask, specifically, what brand of mineral water we serve at the restaurant because they love it so much. So, I guess that's kind of high praise, isn't it?"

"We've gotten a lot of new converts to our mineral water from local businesses like David's," Mr. Omen said. "We've been quite fortunate that the local business community has been so supportive and eager to carry our product."

Mr. Hanson continued, "Since I opened my business sixteen years ago, and I can honestly say that it's one of the best decisions I ever made. Not only does the water taste great, but it also comes with a great story, and we're supporting another local business instead of a big corporation." He smiled and pointed at Mr. Omen. "There's a lot to be said for supporting your friends, am I right?"

"I couldn't agree more," Mr. Omen replied.

Anthony nodded. "My mom and dad like to support small, local businesses, too," he said. "They try not to shop at chain stores if they can find a small business that's owned by someone who lives in the community. I guess it's kind of like what you say about supporting friends."

"Exactly," Mr. Hanson said earnestly, nodding his head along with Anthony's comment. "There's a lot to be said for supporting your friends. I've got some good ones in this here community, and they probably agree with me when I say that this mineral water is as good as it gets."

"How many other businesses carry your water, Mr. Omen?" Anthony asked, looking interested.

"Quite a few," Mr. Omen replied. "We've got a small distribution channel right now, but we're growing all the time. We even have options for businesses who want to use our mineral water under their own private label. We can provide custom labels with their logo, and we're even able to customize the labeling for some special occasions. For example, I know one events venue that serves our water at weddings and other special events. We have a lot of businesses who really like the idea of doing something for their customers that's a little different than what everyone else is doing."

"That's great," Anthony said. "I bet people like that."

"Indeed, they do!" Mr. Omen agreed, nodding enthusiastically. "But the bottom line is this: we've got a fantastic product here in Gold Pyramid Water, and we want to share it with as many people as possible."

"Oh, absolutely," Mr. Hanson replied enthusiastically, nodding his head in agreement. "We're in it for the long haul here."

As Anthony listened to the two men talk about their product, he began to realize just how passionate they were about Gold Pyramid Water. It was obvious both truly cared about their customers, and they wanted to share their products with as many people as possible. There was something really special about this product, and it seemed like everyone who tried it really loved it. He even found himself being more interested in the pyramid house and its story than he ever anticipated.

For the first time in a long time, Anthony felt engaged in the conversation and thought how exciting it was to hear the story of a local business in his community. He couldn't wait to share what he'd learned with his parents and see if they liked Gold Pyramid Water as much as he did.

Yolanda Fierro

Chapter Four

"Are you ready to have your world shaken up a bit?" Paul Omen asked with a twinkle in his eye.

Anthony gave him a patient look. The kind of look you'd give a parent when you're a teenager and your parents are trying to get you hyped up for something that you know is going to be lame. He sighed, seeing as he was stuck in the middle of nowhere with his family because his family wanted to experience the local tourist attraction. "What do you mean?"

"I have stories to tell you. Stories of how the powers of the pyramids have detected and healed lost energies.

"Really?" Anthony raised an eyebrow skeptically. "And how do you know all this?"

"I've actually known people who have benefited from pyramid power. I know people who have been healed of ailments because of the pyramids." Omen said, his voice full of excitement. "It's truly amazing what these pyramids can do!"

Yolanda Fierro

Anthony shrugged. "You seem pretty convinced in the power of pyramids but I'm not one to go in for all this 'woo woo' stuff."

"It's not just woo woo, out there kind of stuff! It's science, my friend. Science that can't be ignored. For example, did you know that a human being's body only accounts for one tenth of their total energy? The mind account for the other 99% of total energy. Pyramid healing targets a deeper level of healing. Soul-level healing, really. It targets the soul and lifts it up, helping us to get rid of all kinds of negative vibes."

Anthony chuckled. "Sounds like meditation to me."

Paul Omen sat back in his chair, and steepled his fingers together thoughtfully. "Well, yes, it's true that meditation can help with healing, and people have meditated inside pyramid structures for centuries, but the pyramids do so much more than just helping you get a grip on your mind for the day. They actually target and heal on a deeper level."

Anthony shook his head. "I don't know about all this. I've never heard anything about all of this before. You said you know people who have been healed from something just from being in pyramids?"

"I do. In fact, I know of four specific people around this area that I'd love to tell you about. And all of them bolstered their natural healing process through pyramid healing. Want to hear about them?"

"Yeah, sure. What do I have to lose?" Anthony was more

than a little curious now. He had never heard of anyone who had been healed by the healing powers of pyramids, but he thought it was worth at least hearing these stories before making up his mind.

"Maybe your skepticism, for one." Paul chuckled. "So first up is Sarah, who has had a history of chronic pain and illness for years. Ever since she was young, she's suffered from all kinds of health problems, including migraines and arthritis."

"Why is that?" Anthony asked. "Was there something in her past? Did she have some trauma that could have contributed to all of this?"

"No, it's not what you think. She actually has a pretty happy, stable family life. But she wasn't always so healthy. Before her husband and she decided to visit our pyramid, she had been suffering from arthritis for several years, and it had gotten to the point where she could barely move around. She was in constant pain and didn't really feel comfortable living her life because of it."

"That's terrible," Anthony said. "So how did this pyramid help?"

"Well, through consistent meditation inside the pyramid structure for about six months, Sarah saw a dramatic improvement in her health. The chronic pain started to fade, and she was able to move around much more easily. She was so happy with the results that she recommended it to all of her friends who have been struggling with their own illnesses."

Anthony shook his head in amazement. "I don't know how you can be so convinced by this stuff, but it's really fascinating. Tell me about the next person you know who has healed from being in the pyramid."

"Okay, well here's Melissa," Paul said. "She is a yoga teacher, and she actually lives in the same town as Sarah. She used to suffer from deep bouts of depression, but after spending dedicated time inside the pyramid for several months, she reported a huge improvement in her mental health. She said that she felt more alive and energized than ever before."

"Wow," Anthony said. "That sounds incredible."

"Yeah, it definitely is," Paul agreed. "Next is Louis. He had severe migraines. Like Sarah, he used to go through bouts of chronic pain from these migraines. He didn't have a good handle on his life before visiting the pyramid structure, but after several months of meditation in it, Louis saw marked improvement."

Anthony suppressed a grin. The older man was nice, in a grandfatherly sort of way, so he didn't want to be rude, but it seemed too good to be true. "Did any of these people see their doctors or start taking new medication around the time they were in the pyramid. Couldn't their miraculous recoveries have just been a result of modern medicine?"

Paul laughed. "I'm the one who knows each of them personally, so you can take it from me that their recoveries weren't by chance. All of them saw noticeable improvements in only a few short months from spending time in the pyramids here." His eyes twinkled. "Think of it. Why can't the power of sacred pyramid geometry, the same kind that helped ancient civilizations build the pyramids, work with modern science and medicine to improve our health and lives?"

Anthony wasn't sure about any of this, but he was getting more and more intrigued. "Tell me about the other people you know who have had healing effects from being inside these pyramids."

* * * *

"Sure thing," said Paul. "Let me tell you about my friend, Jim. He's a lawyer. Very active guy. Always on the move, always hiking, skiing, working out. He has always been pretty healthy and fit."

"Okay. Sounds like he was in good shape."

"He was. Until the accident."

Anthony was instantly alarmed. "Accident?"

"Yeah, he was out climbing a mountain with some friends, and the next thing he knew, he woke up in the hospital." Anthony could tell that Paul cared deeply about his friend Jim. His voice became pained as he continued to speak. "He told me that all of his friends made it off the mountain just fine, but he had fallen and was knocked unconscious. He also broke his left leg."

"Wow. I'm sorry to hear that," Anthony said. "I broke my leg when I was eight. Right at the beginning of summer vacation, too. Ended up in a cast for two months. Couldn't ride my bike. Couldn't go swimming. It was awful." Anthony shuddered at the memory of being separated from his friends all summer long due to injury.

"I'm glad that you were able to heal fully and quickly," Paul said, offering a kind smile. "Anyway, Jim was really in rough shape after that. His leg so badly broken that the doctors weren't sure whether the leg would ever heal properly."

"That doesn't sound good," Anthony said. "I'm sure your friend Jim is very grateful to be alive and well today, though."

"He is," Paul said. "During his recovery, his doctor recommended meditation as a way to manage the stress and pain. Pretty sound idea, right?"

Anthony nodded. "Yeah. I've heard of that."

"Jim was skeptical of it, but he decided to try because he couldn't bear the thought of being in pain for the rest of his life. He trusted his doctor enough to give meditation a shot and it actually helped him feel better. After weeks in recovery though, Jim wasn't satisfied with just feeling better. He wanted to be back to feeling like his old self. So, I convinced him to start meditating here with our pyramid structures."

"And?" Anthony asked, curious about the outcome.

"After spending time in the pyramids here at my place, his doctors were surprised to learn he was healing twice as fast as a man in his age bracket should heal."

"That's amazing," Anthony said. "What happened next?"

"He continued to improve and actually ended up with a full recovery. His doctor was quite conservative, and still left the cast on a full eight weeks to be sure he was healed. But he couldn't believe how quickly the bones had knit back together." Paul paused for a moment, watching Anthony's face as he processed what he was saying. "In addition to the healing that took place, Jim also felt a noticeable difference in his overall well-being. He told me it was like all of his stress and anxiety just melted away."

"Okay, that's one example. You said you had others?" asked Anthony, eager to learn more.

"Actually, yes," replied Paul. "There's another friend of mine named Kelly. She had insomnia and high blood pressure from stress. She went to every doctor in the area searching for relief. They put her on so many medications, but they didn't seem to be helping, so I suggested that she come here and meditate in our pyramids."

"Did it help?"

"Oh yes," Paul said with a wide grin. "Nothing in her routine or prescriptions changed during that time except for her time meditating in the pyramids structure. She came to our pyramids and meditated a few times a week, just like Jim did. Within weeks, her insomnia vanished completely, and her blood pressure lowered significantly. I'm not a doctor, Anthony, but I do know that Kelly was very thankful for the relief she received here."

Anthony was surprised by Paul's story. "That is incredible," he said. "I have some insomnia issues myself with the migraines. It's awful."

"I thought so," Paul said knowingly. "Then there's my friend, Betsey. She suffers with arthritis. Her hands are swollen and stiff. She could barely hold a pencil, let alone sew or knit like she used to. "I'm sorry to hear that," Anthony said sincerely. "Is there a cure for it?"

Paul nodded and smiled again, as if he was happy to share the good news. "There is," he told him. "Betsy had actually lost hope of finding a cure at all after years of suffering. Finally, she came to me. She had heard about Kelly's recovery and asked if pyramid power might be able to help her. I told her we could try, and she agreed."

"And what happened?" Anthony leaned forward, eager for the details of Betsy's story.

"She went through all the treatments doctors suggested, but nothing was helping," Paul explained. "Along with my other friend's experience, she found relief through pyramids. She slept better and noticed her hands didn't hurt so much. Similarly, a friend of hers, called Tim, was having the same health issue you're currently having."

"Migraines." Anthony said, simply, although he knew there was nothing simple about them.

"Yes, indeed," Paul said with a nod. "Tim used to get migraines daily. He suffered terribly from them, and they impacted every area of his life. Work. Family life. Everything."

Anthony sat there for a moment, processing the story he'd just heard. Paul continued to talk about how powerful these structures were and how they could transform lives in all different ways. He explained how they could help people with physical ailments, mental disorders, and even emotional issues.

"I'm honestly amazed every time I hear stories like these," Anthony said finally. "It's just so incredible."

Paul smiled warmly at Anthony as he listened to him speak. "I know, it is amazing," he agreed. "And the best part is that it's completely natural."

Anthony nodded and sat back in his seat. Even though all of these examples were stories from people Paul knew, he couldn't help but feel hopeful that these pyramids really could offer him some relief from his migraines. Still, he was skeptical. Things that seem too good to be true often are.

Paul sensed his hesitation, and his expression turned serious. "I know you may not want to believe it yet, but I promise that pyramid power really does work," he said earnestly. "I've seen it with my own eyes. If it would help you, I could put you in touch with my friends, and they can share their stories themselves. I know it's hard to believe something like this, but I promise that you won't be disappointed."

Anthony took a deep breath and nodded. He knew he had to at least try it. If these pyramids really could offer him some kind of relief from his migraines, it would be totally worth it. "Okay," he said finally. "I'm willing to talk to them. It would be interesting to hear their stories straight from them."

Paul smiled and placed a hand on Anthony's shoulder. "Great," he said. "I think you're going to see some amazing results very soon." He then stood up from his chair and gestured for Anthony to follow him. Before opening the office door, he paused.

"There's another reason why I built this pyramid house aside from being fascinated with pyramids, and learning about the power they can have," Paul said. "I built it for my family."

"Oh, really?" Anthony asked. He noticed how the older man's expression softened with love as he mentioned his family.

Paul nodded. "Yes. I wanted to build something special for them. I wanted my kids, who are grown now, to know that their dad loved them and wanted to give them the world. I wanted that for my wife, too. This house brings us a lot of happiness," he added with a smile.

"It sounds like you have a really beautiful family,"

Anthony said gently.

Paul nodded again. "I do. And I'm so thankful to have them in my life."

Anthony couldn't help but feel a sense of hope as he listened to Paul talk about his family. He wondered if these pyramids really could help him, like it had other people. It might be worth it to try them out.

Even if it didn't help with his migraines, at least he would be able to hang out with Paul's family and learn about their amazing stories to see if he believed the power of pyramids was more than just a cool story.

He was willing to take the risk of trying it and see what happened. He would just have to wait and see how things played out. For now, he decided to focus on enjoying his time with Paul and learning more about the pyramids and all their incredible powers. He was kind of glad he had come with his family on what he thought was going to be a boring day trip after all. There was something quite magical about it all. After all, it was definitely an experience he wasn't going to forget anytime soon.

Anthony nodded, taking in everything that Melissa had said. He felt a surge of gratitude for Mr. Omen introducing him to so many wonderful and caring people who wanted to see him succeed. On the list of people Mr. Omen gave him to talk to, he had one more person to meet with. A man named Louis.

After leaving Melissa's yoga studio, he sent a message to Louis to set up a meeting. He was the one Anthony was really eager to talk with because Louis also suffered from severe headaches. He wanted to hear what Louis had to say and learn more about how he was able to overcome his migraines.

Yolanda Fierro

Chapter Five

Anthony found Sarah Lewis in her small business office at The Serenity Café, a business she ran with her husband of thirty years. The café was a charming, small restaurant and bakery located on a quaint, old-fashioned street in Galena, Illinois. The exterior of the building was painted in bright shades of blue and teal that complemented the warm yellow shutters that adorned its windows. A quaint wooden sign hung over the entrance to the café, bearing the name in delicate cursive script. Flowers and greenery lined the sidewalk leading up to the entrance, lending an air of peacefulness to the space.

Inside, the café was bustling with activity. Customers chatted and laughed as they enjoyed their meals, while the bakers bustled behind the counter, carefully preparing a variety of sweet treats for them. The space was cozy and inviting, filled with rustic wooden tables and chairs that added to its homey charm. The place was infused with a calm, welcoming vibe that immediately put Anthony at ease.

The wall was adorned with colorful images of local art

and landscapes, which were for sale. There was also a large map of the area on one wall, with pins marking all of the local attractions. Anthony noticed there was a pin right over the location of the pyramid structure in Paul Omen's backyard.

Anthony gazed at the menu on the wall behind the bakery counter. There were all sorts of breakfast sandwiches from simple classics like eggs and bacon to more elaborate creations featuring unique, gourmet ingredients. The signs promised the freshest ingredients from local farmers, and—from the smell of things—he knew Sarah's café would beat the local chain store with high priced coffee and generic quick breads any day of the week.

In the bakery case, itself, were sweet pastries and cakes, rich pies and tarts, gooey cinnamon rolls and sticky buns, English scones—with clotted cream and strawberry preserves—and muffins, and dozens of other tasty offerings. Each treat was artfully arranged for display, and the warm aromas wafting through the air tantalized Anthony's senses.

The woman behind the bakery counter smiled at Anthony and politely said, "Can I help you with anything, today?"

"Hi," said Anthony. "I'm here to meet Sarah Lewis. Paul Omen sent me."

The woman smiled. "Oh yes, of course. We've been expecting you. Sarah told me to offer anything you like for breakfast, on the house."

"Really?"

"Of course."

Anthony scanned the menu again with more interest and did the same with the breakfast items in the bakery case. "Well, I'll definitely take you up on that offer. I'm starving!"

The woman chuckled. "Sarah thought you might be," she said with a wink. "Order what you like and take a seat at one of the tables. Sarah will be with you in a few minutes."

He settled on a breakfast sandwich with eggs, bacon, and gouda, a cup of black coffee, and a gooey cinnamon roll that was almost as big as his little sister's head.

He took a seat at one of the tables while he waited, glancing around at all of the other patrons.

The breakfast sandwich was delicious. The croissant was buttery and flakey, and the bacon was crisp, just the way he liked it. Anthony savored every bite of it as he took in all of the sights and sounds around him. Customers came and went as they enjoyed their meals or picked up sweets from the bakery case.

Sarah emerged from the back room just a few moments later, with her long blonde hair pulled up in a loose ponytail on top of her head. She stopped for a moment, on her way to sit with Anthony, and enthusiastically recommended a few dishes to a new customer. Anthony noticed how healthy and vibrant she looked—a far cry from the frail and listless woman Paul Omen had described, who had come to the pyramid structure for healing.

"Anthony?" she asked expectantly as she approached, a wide smile on her face.

"Yep. That's me. Hi Sarah," replied Anthony, rising from his seat and extending his hand. If there was one thing his parents had always insisted on, it was good manners and friendly etiquette. "It's great to meet you."

"Likewise!" chirped Sarah, shaking his hand in return. "Please, sit. I like to get an early start on my day, so I made sure to get here early. I don't like wasting time. It's the most precious commodity we have."

Anthony nodded in agreement as he lowered himself back into his seat. "My dad says the same thing," he said with a smile of his own. "Oh, and thanks for breakfast. It's really delicious."

Sarah smiled again and took a seat across from him. "You're welcome, of course. I'm so glad you could meet this morning." She paused for a moment, then continued with a more serious expression on her face. "I want to thank you for taking up Omen's challenge and coming here to meet me. I know he can be persuasive about the powers of the pyramid. He's hard to resist, isn't he?"

"Yeah, he can be. I'd go so far as to describe Mr. Omen as adamant. Dogged."

"And unrelenting and opinionated. You know what else he is? He's right." She placed her hand over his and squeezed. "He's right about the power of the pyramids."

Anthony chuckled, then gave her a sheepish grin. "I have to admit, it's really hard to believe in all the hocus pocus stuff he talks about."

"I can understand that," Sarah said with understanding in her eyes. "But it's not just wishful thinking or pie-in-the-sky thinking. There's something special about the pyramid, Anthony, and I'm not talking about just how it makes you feel when you're inside of it. There are a lot of people who have experienced healing and rejuvenation, like I have, because of the pyramid. I know it sounds crazy, but it's true."

Anthony peered at Sarah with a newfound curiosity and interest. He had to admit that she did look incredibly healthy, and not just because of her tan skin or toned physique. There was an inner glow about her—a sense of peace and serenity that he hadn't expected.

"Okay," he said slowly, still not ready to be convinced as he leaned forward intently. "I've heard anecdotes from Mr. Omen, but I'd really like to hear your story directly from you."

Sarah nodded and gave his hand another squeeze. "Not a problem. I love telling people about my experience.

"I had a severe case of rheumatoid arthritis for a long time. Years, in fact." She paused, seemingly lost in memories of the past. Anthony could see the shadow of pain as it flit across her face like a cloud passing over the sun.

"It was debilitating," she continued after a moment, as though reading his expression. "I couldn't even walk without assistance. At my worst, I couldn't get out of bed without help—the joint pain and swelling had become so bad that even a can or a walker wasn't enough support. I was in a wheelchair. It got to the point where even simple tasks like getting dressed or making breakfast in the morning were impossible."

"Wow," Anthony said, softly. "I'm so sorry that happened to you."

"Thank you," she said, with a slight smile. "I was so jaded and depressed about my condition that I didn't believe anything could help me."

"Mr. Omen mentioned that you had some other health problems, like migraines?"

She nodded. "I'd been diagnosed with migraines before I met my husband. The started about as chronic headaches without a pain trigger to avoid. Nothing too bad, at first. But then they started getting worse and I couldn't get them to subside. The medication didn't work very well, either."

"That's so terrible," Anthony said with real sympathy in his voice. He remembered the last time he'd gotten a migraine—it had been awful enough that he'd gone to urgent care for treatment. His mom had to drive him because it was too painful to even see straight and being in direct sunlight was like getting jabbed with a thousand needles.

"A few years ago, my migraines went from bad to worse. I thought I was going to go blind. The pain got so bad that

Passion of the Gold Pyramid

the only way I could get relief was to lie flat on the floor in a completely dark room. And sometimes the pain was so bad, it felt like I was having an out-of- body experience."

"That sounds awful," Anthony said again, wincing at the thought of those headaches.

Sarah smiled in a wry, self-deprecating way that made him like her even more. "Let's take a walk. I want to show you something."

Anthony followed Sarah to the back of the restaurant, up a flight of stairs and into her office. The walls were decorated with pictures of smiling people—family and friends, Anthony assumed—and the furniture was arranged to be comfortable and inviting. The desk looked out over a wide deck that had a view of the Fox River.

"Nice view," Anthony said appreciatively, walking over to the window and gazing out at the view. "I've spent every summer I can remember on that river. Well, except when I had a massive migraine, that is."

Sarah sat down at her laptop and tapped on the keyboard several times. "Come over here. Take a look at these pictures." She turned the laptop toward him.

Anthony walked over to her desk and saw that she had opened a folder of pictures. They were all of a very young Sarah, looking fit and healthy. There was one where she was wearing a bikini and sunning herself on a beach in some exotic locale. She had an intense tan and gleaming white teeth. "This is me before the arthritis hit."

Anthony smiled at her. "You look great."

"Looked great. Past tense. Take a look at these photographs." Sarah clicked on another folder and more pictures. She set the display to run a slideshow.

Slowly, the young, vibrant Sarah was replaced by an increasingly frail woman who seemed to be in constant pain. In each photograph, the smile lines were deeper and more pronounced. The arthritic hands curled in on themselves and it looked painful for her to clutch a cane or a walker.

"Wow," Anthony said softly, horrified by the progression of Sarah's disease. "I can't believe how much the arthritis changed you."

"Believe it. Arthritis is a cruel disease that can rob you of everything you have. Everything you love. The pain is debilitating, and the medications can often make you feel like a zombie. It's hard to stay positive when all you're able to do is just . . .manage pain."

"I can't even imagine," Anthony said, shaking his head.

"Now look at this." She clicked over to another set of pictures. "I'm not sure if you noticed, but I stopped taking my medications. There was a point when I couldn't take the side effects anymore—they were really bad. So instead of just feeling like crap all the time, I decided to try something new. Something unconventional."

"The pyramid meditation?" Anthony asked curiously.

"Yes. I had run into my old friend, Paul, and he noticed how much pain I was in. He told me that he had been dealing with a lot of pain issues himself, but his problems were almost completely gone because of the pyramids. He said I just had to have faith and trust that this could work for me."

"You never thought about giving it up?" Anthony asked her frankly.

She sighed. "Of course, I did. When you're in that much pain, it's really hard to believe that anything could help. But I didn't want to be in pain any longer, and Paul was adamant the pyramid power could help, so I told Paul that I would give it a try. It took time. Results were a bit slow, at first. But then, finally, I started seeing improvement."

Anthony considered her for a long moment. "Did you start any new treatments or medications around the time you started the meditations in the pyramid structure?"

Sarah shook her head. "No, nothing else changed, other than the meditation in the pyramid. I still saw my doctor regularly, and she was amazed. She couldn't explain why everything was so much better without making medication changes."

"Are you still meditating in a pyramid?" Anthony asked her.

She smiled at him. "Every day. I don't even think about it anymore, because the improvements have been so dramatic." She held out her now-fully-functioning hands and wrists.

Her smile was wide and radiant. "I feel amazing and it's all thanks to the power of the pyramid. It's really a miracle."

Anthony nodded, his eyes shining with admiration for her strength and her faith. "I have to admit, your story is impressive."

He knew firsthand just how difficult it could be to live with arthritis and chronic pain, but Sarah had managed to find healing in an unexpected way.

A sudden, sharp pain in his temples made Anthony wince, and he reached up to rub the spot. Sarah peered at him, concerned. "You okay, kiddo?"

Anthony shook his head slightly. "Headache. I hope it doesn't get worse, though. Otherwise, I won't be able to stick around for long."

"You can stay as long as you need," Sarah assured him gently.

Anthony sighed and leaned back against the cushions of the couch that sat in front of Sarah's pyramid structure. "I just wish there was a better solution than medication. It makes me so groggy. Sometimes, I have to take enough to knock me out."

"I think you know exactly what you should do. You should try meditating in the pyramid," Sarah told him gently. "I'm sure it would help with your migraines. In fact, I have a pyramid out on the corner of the deck. Would you like to try it for a few minutes?"

Anthony peered at her through squinted eyes, considering her offer. He knew that his options for reducing or even eliminating his migraines were limited, and the pain in his head was getting worse. He was willing to try anything.

"Okay," he told her finally. "Let's give it a shot."

Anthony followed Sarah outside onto the deck, where she gestured toward the pyramid structure in the East corner of the deck.

Sarah smiled gently at Anthony. "The power of the pyramid is real, and I wouldn't hesitate to recommend it to anyone struggling with pain." She reached out and gave his hand a squeeze before helping him get settled inside the pyramid. "It's all about faith and believing that something can work for you, even when the odds are stacked against you. So, the big question, now, is this: Are you ready for your life to change?"

Yolanda Fierro

Chapter Six

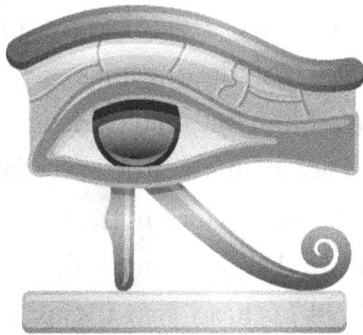

*A*nthony arrived early the next morning at the pyramid house. As the tour guide, Harold, approached the door, he smiled widely in recognition as he peered through the glass. The door swung open and Harold, the tour guide, greeted him like an old friend. "Ah, Anthony! It's good to see you again! It's a little early for tours. We don't normally start at seven o'clock in the morning." He laughed. "Your mom, dad, and sister aren't with you this time?"

Anthony shook his head. "No, it's just me. Mr. Omen said I could come by anytime I like and meditate in the pyramid. I've been having a lot of headaches lately and he thought it might help."

Harold chuckled. "Such a kind man, Mr. Omen! Yes, the pyramid has some healing effects on people. The more time you spend in here, the better you feel." He ushered Anthony in, then turned and closed the door behind them.

Anthony pulled out his phone to check the time. He had plenty of time to mediate before meeting with Mr. Omen's friend, Melissa, and it was always a good thing to get some quiet meditation in before he started work for the day. He followed Harold to a small, quiet room upstairs.

In the center of the room there was a large pyramid made from copper. It was twice as tall as he was, in the center of the structure were several cushions and pillows that he could use to sit or lay upon.

The room itself was dark and peaceful. Harold directed him to a small desk against the wall with a laptop and, when he tapped the touch screen, a list of meditation playlists displayed. Anthony selected a nature-based playlist—one with ocean sounds beneath the music—and pressed play. It was a soft, ambient sound that immediately sounded soothing.

Harold lit candles, strategically placed around the room, and their gentle light flickered over the copper pipes of the pyramid structure. The scent was heady and earthy, like sandalwood or patchouli. He breathed in deeply, letting the soothing aroma calm his mind.

"Looks like you're set, young man," Harold said. "I'll leave you to it. Just let me know when you leave so I'll know you're okay." He smiled and exited the room, closing the door behind him.

Anthony looked around the room, taking in the copper structure and the Egyptian decor around the room. There were small statues of Bastet, the cat goddess, Horus, and other Egyptian deities lining the floor-to-ceiling bookshelves on two of the walls.

Anthony took off his shoes and slid into the pyramid, sitting down in a cross-legged pose on one of the cushions. The metal was cool beneath his skin, and he found himself shivering slightly as he settled in to begin his meditation. He sat cross-legged on a pillow in front of the pyramid, closing his eyes to focus on his breath. Anthony had woken with a headache, and his head still hurt. He hoped that the meditation would bring some relief, and the pyramid would help the calming effects last longer.

The music was soft and soothing, and Anthony felt his breath begin to even out as he sank deeper into the meditative state. The soothing sounds of the ocean filled his head, washing over him like a warm tide. He breathed deeply and let himself relax. Slowly, the sounds of nature faded away until he was left with only the sound of his own heartbeat. When a friend's mom first introduced meditation to him and his a few of his other friends, it felt awkward. He thought it was boring and pointless and didn't really understand what it was meant to do.

The first time he tried it, Anthony couldn't keep his mind from wandering. He thought of all the things he had to do that day, and all of the things he'd done earlier that day. He started a mental to-do list in his head and realized he hadn't eaten breakfast yet. All of these things pulled at his mind, reminding him that they were things he had to do or fix. He felt like he was never doing enough because all of those thoughts kept popping up and distracting him and his mind raced with thoughts about what might or might not happen in the future.

But slowly, over time, when his mind wandered, he learned he could slowly bring the focus back to meditation by just focusing on his breathing. Or by repeating a simple word like 'feeling' or 'thinking' to acknowledge where his mind had wandered, and then pulling it back to the present moment.

Eventually, Anthony was able to fully immerse himself in the present moment and he could feel his mind relax as he let go of all those extraneous thoughts that seemed so important on the surface but didn't really matter in the grand scheme of things.

As Anthony sat in the pyramid, focusing on his breathing and shutting off his thoughts, he felt his body slowly letting go of the tension that had built up. He was able to relax and feel more grounded in the present moment. The headache began to slowly subside. First, it lessened in intensity, but then it slowly diminished to a dull throb. Anthony felt a sense of calm settling over him as his body relaxed and his mood was lighter, too.

Then he felt a warmth through is upper body, radiating from his chest outward to his arms and shoulders. Anthony had never felt anything like this before, yet it seemed so familiar somehow, like something he had experienced in another lifetime. It almost felt like a memory from long ago that was still tangled up in his mind somewhere. As the warmth spread through him, Anthony began to feel a lightness that surprised him.

He felt more awake but less anxious, and at that moment, the sounds around him—the music, the faint buzzing of a nearby light, and the quiet chirping of birds outside—seemed quieter than before. As the minutes passed, his thoughts faded away until he was no longer aware of himself or his surroundings. He lost himself in the depths of his mind and the peace of being in the present moment. It was easy to forget about everything else when he was focused on meditating, yet Anthony knew that this meditation would help him stay grounded throughout the day.

With a calm mind and relaxed body, he could face whatever challenges came his way with ease. He wondered whether it was really the effects of the pyramid's power to focus the mind and calm the body, or just the placebo effect of wondering whether the pyramids could help. He had to admit, in all the months he had been meditating to help his headaches, he hadn't before felt anything like what he was experiencing in the pyramid.

There was a knock on the door and Harold tentatively poked his head in. "Everything okay in here?" he asked.

"I'm fine," Anthony said, his voice slightly distant as he

continued to focus on his breathing. "Just need a few more minutes."

Harold nodded. "Ok. Just checking. You've been in here almost ninety minutes, and I just wanted to make sure you didn't need anything."

"Ninety minutes? You're kidding," Anthony said, surprised. He looked at the clock on the wall. It was nearing 10:30am and he had meant to only spend a thirty- or forty-five-minutes meditating. "Wow. I've never gone this long before."

Harold shrugged. "It's just you and the pyramid today, so I thought maybe you were getting a little more out of it than usual." He gave Anthony a small smile before leaving him alone to finish his meditation session.

Anthony slowly got up from his meditation seat and headed back towards the living room. As he walked out of the pyramid, he felt more relaxed than ever before. He wondered if this was how the ancient Egyptians had been able to focus their minds, meditate, and find true peace within themselves without any modern conveniences like electronics or medical treatments. As he extinguished the candles and turned off the music, he took a last look around the room.

His headache felt better. It wasn't as intense as it had been when he first arrived. He was feeling less anxious than he had before, and there was definitely a lightness to his body and mind that made him feel more grounded in the present moment. He checked his watch and realized he had just enough time to make his meeting with Melissa at the yoga studio.

Traffic was light, so Anthony arrived at Melissa's studio with time to spare. He walked into the reception area and greeted the lady behind the welcome desk. She was a young woman with a sweet smile and big brown eyes, and she immediately welcomed him with a warm hello.

"Hello, I'm Anthony. I'm here to see Melissa."

"Oh sure," she said. "She's walking through a new program she created. She asked me to bring you in when you arrived." She smiled again ushered him into the yoga room.

The room was light and airy with a large window overlooking the town square. There was yoga-inspired artwork scattered around the space—several sculptures of lotus flowers, and paintings of yoga poses. A soothing scent wafted through the air and a calming instrumental music played in the background. Melissa stood on a yoga mat in the middle of the room, dressed in comfy workout clothes with her long brown hair pulled into a messy bun.

She was in the middle of a pose and looked up when Anthony walked in, giving him a radiant smile.

"Hey there, Anthony. I'm glad you could make it." She came out of her pose and walked over to greet him. "I was just finishing up working on a new yoga routine for my students."

Anthony smiled back. It was hard not to when he was

around someone as warm and friendly as Melissa. She had a radiant energy, and he could feel the positive vibes oozing off of her.

"I'm glad to be here," Anthony replied. "Thank you for having me."

Melissa shrugged. "It's my pleasure. Would you like some green tea? I can brew some for us."

"Sure. That would be great."

Anthony followed Melissa to an anteroom behind the workout space. There was a counter with an electric kettle, tea bags, and a few mugs. There was a small table with four chairs in the middle of the room, and in one corner there were two comfortable cushy chairs with a small table between them. Plants and flowers were set on each windowsill and the smell of fresh jasmine filled the air.

"This space is new," Melissa said, gesturing towards the room. "I've started hosting special events for my students in this part of my studio. I'm hoping to have a teatime one evening per week, so a few of my clients can hang out and relax."

"It's a really nice space. My mom loves yoga, and she might like to come and hang out here, too."

"I'm glad to hear that," Melissa said. She filled up the kettle with water and turned it on while Anthony sat down in one of the cushy chairs. She plopped tea bags into two mugs and then poured in hot water from the kettle when it was done boiling.

"So, Paul told me all about your conversations with him, and the meditation you've been doing on the pyramid," Melissa said. "I'm so excited to hear about it."

Anthony smiled and took a sip of his tea. He let a contented sigh and leaned back in his chair. "I had my first meditation in one of the pyramids at his house this morning. It was amazing, really. Powerful in ways I wasn't expecting. "I'm not surprised," Melissa replied. "The pyramid is amazing, and it seems like a special connection has formed between you and Paul. He's really special, by the way." She smiled, and Anthony felt a certain warmth in his chest, like he was glowing from the inside. "He is so giving and honest. I feel like he's one of the most authentic people I've ever met. Everyone I know who has met him, become friends with him, has felt the same thing. I just adore him."

"I'm learning that about him," Anthony said. "Everyone I've met has told me story after story about how he has helped them, supported them, and lifted them up. And his family adores him, too. He's the kind of person who seems to bring out the best in those around him."

"That's because he has a heart of gold and is fully present with everyone," Melissa said. "I'm so glad that you've connected with him. I hope you can continue to spend time together and get to know each other better." She took a sip of her tea and beamed at Anthony. "He certainly helped me when I needed it the most."

"I'd love to hear your story. How did he help you? And how has the pyramid powers helped you?" Anthony asked. His curiosity was piqued, and he felt hopeful that there might be some insight in Melissa's words that would help him better understand what was happening to him.

"Well," Melissa said, "for as long as I can remember, I've been dealing with bouts of intense depression. I've tried therapy and medication, but nothing seemed to help. My doctor told me that it was just part of my personality, that this would be something I would have to learn to live with for the rest of my life."

"Sounds like what my doctors told me about my migraines."

"But Paul helped change all of that after we met. He told me all about the powers of the pyramid, and how they could help to heal my depression. He suggested I spend dedicated time meditating in the pyramid, and so I did. And you know what happened?"

"What?" Anthony asked, his eyes wide with fascination as he listened intently to Melissa's story.

"It took time, but after a few months, I began to feel better. I started coming out of the pyramid session feeling like a completely new person," she said. "It was the first time in my life that I had experienced something like that, and it was absolutely miraculous. It was like it supercharged my ability to manage my depression. I still take medication, of course, but I also have a tool that I can use to fight my depression."

"This sounds really promising. How is your depression now?" Anthony asked. "Are the effects of meditating in the pyramid long-lasting?

"It's like any meditative practice, you have to be consistent with it. It's not a one and done kind of thing." Melissa said. She smiled and gazed off into the distance for a moment, as if she was conjuring up memories of her time in the pyramid. "I have my ups and downs just like everyone else, but I am so grateful that Paul introduced me to this amazing power that I can harness whenever I need to. It's a constant reminder of how powerful and interconnected we all are, and how much love we have to share with those around us. Now that you've started to meditate in the pyramid, don't give up on it. Give it time to work. You'll feel better soon."

Chapter Seven

The alarm buzzed loudly, jolting Anthony out of a deep sleep. He groaned and turned over, trying to get comfortable again. But the noise wouldn't stop.

"Ugh," he muttered as he reached out for his phone on the nightstand. As his fingers fumbled with the screen, he wished that he could go back to sleep.

But the alarm kept ringing, and there was no ignoring it. Why had he set the alarm so early again? He didn't need to be at work until early afternoon. With a sigh, Anthony finally managed to silence the alarm and sat up blearily as he rubbed his eyes. He glanced over at the clock on the nightstand—7:30am. What was he thinking when he set that alarm so early?

His meeting with Louis, Paul Omen's friend and a fellow migraine sufferer, came to mind. He remembered it was scheduled for eight o'clock. He only less than an hour to get ready and get on the road, and it was a bit of a drive to where Louis lived, so there was no time to waste.

With a groan, Anthony dragged himself out of bed and

headed into the bathroom to take a shower. As he stood under the warm spray of water and let it wash over him, he began to feel more awake. He paused to take stock of the headache situation. Most mornings, he woke with a vaguely dull ache in his temples that had a 50/50 chance of spreading to the rest of his head. Right now, he was actually feeling pretty good. Maybe the meditation sessions at the pyramid house were actually working.

Anthony finished his shower and got dressed. As he made his way out to the kitchen to fix himself some breakfast, he smelled coffee, bacon, eggs, and toast. His mom must already be up, he thought with a smile. She was always the first one to wake up and make breakfast on weekends.

"Morning, sweetie," she said cheerfully as Anthony walked into the kitchen. "How'd you sleep?"

"Not too bad," he replied, grabbing a plate from the cupboard and loading it up with breakfast. "And you?"

"Pretty good, actually," she said, pouring him a cup of coffee. "It's so nice to have a weekend where I don't have to wake up early for work." She glanced at the clock on the microwave. "You're up early for a Saturday." She brushed the hair out of his eyes and gave him a quick kiss on the top of the head—a morning ritual between them since he was a kid. "I thought you didn't have to work until later."

"I don't," Anthony said, taking a sip of coffee. "I'm actually going to meet with a guy named Louis today."

"Louis." She frowned, thinking. "A new friend, dear? Have I met him?"

"Not a friend of mine, a friend of Paul Omen's."

Mom paused as she handed a plate to Anthony's dad, who had just entered the room, and then filled a plate for his little sister, Jenna. "Omen. You mean the man who owns the house shaped like a pyramid?"

Anthony piled his scrambled eggs on top of his toast and squirted a generous portion of ketchup on them. Then he selected the crispiest pieces of bacon on the plate—the almost burnt ones he loved so much—and added them on top of the eggs before closing the sandwich with the last piece of toast. "Yep. That's the guy."

Dad paused in the middle of chewing a mouthful of bacon. "How did you meet him?"

"I've haven't exactly met him yet. Today will be the first time. I've been spending time with Mr. Omen. He's told me a lot of behind-the-scenes things about the pyramid house and the mineral water they bottle." He took a big bite of his breakfast sandwich, chewed and swallowed. "I guess Louis suffers from migraines, like I do, and the meditation he does in a pyramid structure—like the ones at Mr. Omen's house—help make them better."

Jenna bounded into the kitchen like an over excited puppy. She was a morning person, like Mom. Anthony and his dad were the opposite, preferring to sleep in and wake up slowly. "Morning." She sat down and grabbed a piece of toast from the plate in front of her and stared at the eggs. "Can't I have cereal instead? I don't like eggs."

"They're much better with ketchup on them," Anthony said patiently. He always knew how to cajole his sister into almost anything. He showed her his breakfast sandwich. "See? Look at how good these eggs are."

Jenna poked her fork into her eggs and then plopped them onto her toast. She squirted ketchup on them and added bacon like her big brother. She took a bite and chewed thoughtfully. "Needs cheese," she said.

"Alright," Anthony replied, a grin spreading across his face. "How about if I slice some cheese onto your eggs?"

Mom laughed and took a seat at the table. "She's definitely your sister." She looked at Anthony closely, her eyes narrowing slightly in suspicion. "He meditates in a pyramid structure, and it helps his migraines?"

Anthony fetched some cheddar cheese from the fridge and sliced a couple of large pieces that he laid on top of Jenna's eggs. She took a bite and her eyes widened in pleasure before she quickly finished the rest of her breakfast.

Anthony replied, "That's what Mr. Omen says. "I've met a few people who know Mr. Omen, and he suggested the power behind the pyramids would help them with different health problems. Like Melissa who suffers from depression, and Sarah who has arthritis. Their stories are pretty compelling."

"You sound like you're starting to believe him," Dad said, taking a sip of coffee. "I've never heard about pyramids having mystical powers before."

"I was skeptical," Anthony admitted. "But those people he's helped are real. That has to mean something." He paused. "And I've done some meditating in a pyramid at Mr. Omen's over the past week or two and, I have to say . . . today is the first time in a long time that I've woken up without a bit of a headache to start the day."

Mom shook her head, a smile playing at the corners of her mouth. "I think you might be getting a bit sucked in by all this pyramid power talk." She glanced at Anthony's dad, then back at Anthony. "But I'm glad you're feeling better. That's really what matters."

Anthony polished off the rest of his breakfast sandwich and pushed away from the table. "I'm going to head out," he said, standing up. He kissed his mom on the cheek and ruffled his sister's hair. "Keep an eye on Dad so he doesn't eat all the bacon."

***

Anthony drove to Louis's house, which was on a farm about twenty miles outside of town. As he pulled onto the driveway, Anthony felt a surge of excitement at the prospect of meeting another friend of Mr. Omen's for the first time. He hoped Louis would be as friendly and welcoming as Mr.

Omen had been—and Sarah and Melissa—and that he'd have more information about the healing powers of the pyramids.

As Anthony stepped out of his car and made his way towards the front door, an apricot-colored shaggy dog with a coat like a velour blanket ran at him and jumped up, its paws almost knocked Anthony to the ground. He laughed and petted the dog, steadying himself on his feet again. Then he knocked on the door.

A burly man with a beard and clear blue eyes answered the door with a big grin on his face. He was dressed in shorts and a t-shirt, his feet bare. "You must be Anthony! I see you've met Charlie."

"Yeah, he's a pretty friendly dog," Anthony said.

"He's the sweetest thing," Louis agreed. He led Anthony into a cozy living room and sat down on a couch. "He just wants to hog all the attention from everyone he meets. Anyway, I'm Louis. Want some coffee?"

"Sure. I could always use more coffee. Caffeine helps with the migraines unless they get too bad." Anthony nodded eagerly and followed Louis inside.

Louis gestured to a large, dark wooden table in the corner of a sunny breakfast room just off the kitchen. The space was decorated in a calming shade of blue and filled with light, making it feel bright and inviting. "Grab a seat in there, I'll bring the coffee in."

Anthony sat down at the table and glanced out the nearby

window which overlooked a pond with a weeping willow tree. He closed his eyes and took in a deep breath, trying to center himself before continuing the conversation with Louis. He opened them again, still feeling relaxed and calm from the view of the pond and peaceful sounds of nature drifting in through the open window.

After about few minutes, Louis brought in a tray of mugs and a carafe filled with steaming hot coffee. He set it down on the table and poured a mug for Anthony, then settled into a chair across from him.

"I heard about your migraines," Louis said after taking a sip of his own coffee. "I hear you have a lot going on at work, and you're in college, too, right? That's quite an intense schedule to keep. Especially when you're plagued by migraines. I'm sorry."

Anthony sighed and nodded. "I'm off school for the summer and working full-time. But yeah, it's tough when school is in full swing and I'm still working. I mean, I don't think anyone would be able to handle all that pressure without any problems. But the migraines make it so much harder to handle."

Louis gave him an empathetic look, one that only someone with severe migraines would fully understand. "I know, it can be really hard to deal with, especially when you have so much else going on in your life. It's lucky you're getting some time off this summer though."

"Yeah, I'm definitely glad for the break," Anthony agreed. He took a sip of coffee and felt some of the tension in his body begin to melt away. "I have to admit, when I first talked with Mr. Omen about the pyramids, I thought it was just a crazy story. But . . ." He trailed off, thinking about his recent experiences inside the pyramid and all he was starting to learn from it. "But I've met with several people he helped over the years, and they've all shared amazing stories about the healing powers of the pyramids. It's pretty incredible."

"It really is," Louis agreed. He leaned in to share a conspiratorial smile. "You know, when Paul first told me about pyramid power, I thought it was just a bunch of hooey. I've had trouble with migraines for decades and tried just about everything to ease it—yoga, acupuncture, massage therapy, medication—nothing helped much. But Paul has a way of changing the lives of people he cares about for the better, and he challenged me to give meditating inside a pyramid a try. When he said, 'what do you have to lose except the pain,' I realized he was right. So, I gave it a try."

Anthony couldn't help but feel a bit of envy for Louis as he listened to him talk about his experiences with pyramid power. "So, you really think this is an effective treatment? I mean, long term?"

"Absolutely," Louis said firmly. "I've been meditating in the pyramid twice a day and feeling better every day. It's incredible to feel that much relief from pain after so many years of struggling with it." He smiled at Anthony as he continued. "I've been meditating inside the pyramids for several years now, and I'm living proof that the stories are true. It's improved my health so much, including reducing my migraines to only a couple times a year, and even then, the migraines come back when I get lazy and don't spend enough meditation time in the pyramid."

"Tell me more about the migraines you have," Anthony asked. "How bad are they? What symptoms do you experience?"

Louis gave him a thoughtful look. "It depends. Sometimes it's just an ache in my head somewhere, sometimes it feels like a bowling ball is crushing my skull, and other times I get dizzy before the pain comes on." He shrugged helplessly. "I know a bad one is coming when the aura hits and I start to get that feeling of being off balance. The aura is the warning system for me, and when it happens, I know the pain will come soon if I don't do something about it."

Anthony nodded thoughtfully and took another sip of coffee. "Is the aura just visual?"

"Sometimes that's all there is, but sometimes it also involves a tingling sensation on my skin as well," Louis said. He ran his hand over his cheeks, indicating where the tingling sensation would be.

"And have you found anything in particular that seems

to trigger a migraine?" Anthony asked, trying to think of possible factors he could address during their time together in the pyramid. "Everyone experiences migraines differently and there are so many things that someone with migraines might encounter. I had done breathing exercise before, too. But when I started meditating in the pyramids, it was like it amplified all of the preventative and pain management interventions I was already doing. It was so much more effective when I meditated in the pyramids, and I think the pyramid powers help put me into a focused state of mind where I can better focus on my treatments."

"That makes sense," Anthony said. "I've already had some small improvements in my pain symptoms meditating at the pyramids."

Louis gave him an encouraging nod. "A lot of people get relief from the pain using pyramid power," he said. "There's something about the way those pyramids are built—the sacred geometry thing that Paul talks about all the time—that just seems to help people heal and restore their health. I'm so grateful Paul told me about all this, and that I had the chance to try it for myself. It's a wonderful tool for reducing my pain and improving my overall health."

Anthony considered what Louis had said, nodding in agreement. "It does sound like the pyramid has the power to transform people's lives for the better," he said. "I'm curious to see how this helps me heal and feel better. If it works for me, I know a lot of other people who suffer from migraines would be interested in trying pyramid meditation."

"Oh, definitely," Louis said. "The people I've talked to about the pyramids are really excited to try it out for themselves. So, I guess the next question for you, young man is, are you ready to go all in with this pyramid power thing?"

Anthony smiled and nodded. "I think so," he said. "Like Mr. Omen said to you—what have I got to lose except the pain?"

Chapter Eight

nthony arrived at the Gold Pyramid house, excited to meditate in the energy of this mysterious building. As soon as he walked through the front door and his eyes adjusted to the dim light, he noticed the tour guide, Harold, preparing for the first tours of the day.

"Hey, Anthony," Harold said with a warm smile. "How ya doin', kid? Here for some meditation time?"

"Yes, I am," replied Anthony. He looked around the large entryway and observed the beautiful ancient artifacts that adorned the walls. The place had such an incredible energy, he thought. There was just something so magical about it.

After talking with Sarah, Melissa, and Louis about Paul Omen and all of his dreams and plans, Anthony had become even more intrigued by the man who built the Gold Pyramid house. He wanted to know more about what made this guy tick, and he hoped that Harold might have some insight into who Mr. Omen really was.

"Hey, Harold. I was wondering about your impression of Mr. Omen. I've been talking to a lot of people he helped over the years, and they all tell me these amazing stories about him. I'd really like to hear your take on him."

Harold smiled and looked over at Anthony with a knowing look in his eyes. "Paul Omen isn't just an amazing businessman, he's a good friend too," he said, as if reading Anthony's thoughts. "He is truly one of the most wonderful people I know. He has an incredible talent for dreaming big and turning those dreams into reality. When he sees something, he wants, he goes after it with everything he's got."

"Yeah? So, the pyramid house and the spring water business aren't his only projects, then, huh?"

"Oh, no. Mr. Omen is a businessman through and through."

"What else has he done? If he's done a lot of other projects, that's gotta be one of the reasons for his success. I mean, you don't have the money to build something like this without a lot of other achievements behind you." He gestured around the grand room with all its artifacts and the viewpoint down into the aquifer.

"You've got that right. Mr. Omen has had his hands in a lot of things over the years, but I can tell you he's been doing this for around thirty years or so." Harold smiled fondly. "Just think about all of the people who have come and gone through these doors at the Gold Pyramid house over the years. Thousands of visitors have spent time with us. All of them have been touched by Mr. Omen's fantastic vision for this place."

"The man is a building genius. His businesses are all successful, but it's not just the money that drives him. He really cares about big visions and innovations, too."

"Harold is being very kind in his estimation of me." A booming voice echoed down from the top of the pyramid, causing Harold and Anthony to turn their heads to see Paul Omen coming down the stairs. "Good to see you, Anthony. I hear you've taken me up on my offer to use meditate here as often as you want."

"Yeah. It's just as I imagined. It has been incredible, Mr. Omen. Learning about the powers of the pyramid, I mean."

"Well, I'm glad you're enjoyed it here," Mr. Omen replied with a smile. "When you're done with your meditation, come find me. We'll have a little chat to catch up. I've got some ideas I'd like to share with you."

"That sounds great, Mr. Omen," Anthony replied gratefully. He knew that Paul Omen was a man of many talents and accomplishments, but he still found it a little hard to believe that such an accomplished businessman had the time to talk with him about anything at all.

Mr. Omen excused himself to go back to his office, and Harold clapped Anthony on the shoulder before returning to his preparations for the mornings' tours. Anthony headed to his usual meditation spot. Before he was in the room, he could feel the power of the pyramid surrounding him. It felt like calm waves of energy washed through his body, and he knew that the meditation was going to be one of those times when he just didn't want it to end.

Anthony prepared the room. He lit the candles and put on some mood music that helped him relax. He slipped off his shoes and sat beneath the pyramid frame in the middle of the room.

He closed his eyes. Breathed deeply. When he let his mind go still, he could feel the energy of the pyramid around him. He let serenity flood into him as he imagined himself like a sponge soaking up water.

The longer he sat—mind still, heart open—the more tranquility he felt. The frequent migraines he had, so debilitating for years, were all but gone. He felt like he was finally on the path to living a life that was more fulfilling. He could see his own potential.

Anthony's meditation session was over too soon, as it always seemed to be. As he left the pyramid room and headed for Mr. Omen's office, he felt rejuvenated. He was amazed at how refreshed and optimistic he was about his future, and what any new changes might bring for him.

He knew that whatever kind of chat Paul Omen had in mind, he would learn something new—perhaps something he might be able to use for his future—and Anthony was excited about what it might be. At the very least, Mr. Omen's advice would be fresh, interesting, and not something he'd heard a hundred times before.

Anthony knocked on the office door, and Mr. Omen invited him in with a smile. Mr. Omen was seated at his desk—a large solid wood piece, with an executive feel to it— and he gestured toward a chair in front of it.

"So, young Mr. Johnson. You're enjoying your time here at the Pyramid house."

"I'm really enjoying it, Mr. Omen," Anthony replied. "It's been amazing to be able to meditate in such an incredible place with all of this energy around me."

Mr. Omen nodded slowly, then flashed another smile. "I think it's time you called me Paul, don't you?"

Anthony grinned and nodded. "And you can call me Anthony. I think we're on such great terms, it seems silly to keep all that formality between us."

Paul laughed at the joke. "I agree. I always enjoy our chats. You're a bright, interesting young man, Anthony."

"That's very kind of you to say," he replied. "I always felt that way about you, too."

Paul leaned back in his chair and nodded. "You remind me of myself at your age. Smart. Inquisitive. Determined. I think you can do so much with your life, Anthony."

Anthony was more than flattered, he was gobsmacked. It was hard to believe that someone like Paul Omen, who had not only built the incredible Pyramid House, but was also someone with such an impressive business acumen, would see potential in him at all.

Paul paused for a moment and looked at Anthony. "I think it's time you started thinking about what you want."

"Really? You think so? Because now that I've graduated high school, I don't know what to do," he said honestly. "I have a place at Northwestern University in the fall, but I don't know what degree to go for."

"Yes," Paul said. "Have you given it any thought? Your future, I mean."

Anthony thought about it for a moment. He had a lot of hobbies—he'd been interested in photography and film for a long time, but he wasn't sure where that would lead him. He also enjoyed writing and reading a lot, so perhaps journalism or creative writing might be something to consider.

"I have thought about it," Anthony said. "But I'm really not sure what direction I should take. There are so many options." Paused and took a deep breath before letting it out slowly. "I'm just not sure what to do."

Paul smiled at him again, and this time it was a kind, supportive smile rather than in amusement. He leaned forward in his chair, placing his elbows on the desk. "Have you ever heard the phrase: If you love what you do, you'll never work a day in your life?"

"Yeah, my dad says that sometimes. Whenever he does, my grandad always says, 'they call it work for a reason'."

Paul chuckled. "Okay, how about this one. 'Discover what you're good at and what you're passionate about and find a way to make money doing that.' "

Anthony perked up. That was something he had never heard before, but it seemed to be reasonable. He leaned forward, eager to know more. "What exactly do you mean?"

Paul's eyes twinkled with excitement—he clearly loved what he did. "Sometimes, you might be good at something but aren't passionate about it. You'd be able to earn money, maybe even make a successful career from it, but it wouldn't be fulfilling."

Anthony nodded. "That makes sense. I know so many people like that—people who have jobs that aren't really their true passions, and they're miserable because of it."

Paul leaned back into his own chair once more. "But if you match what you're passionate about with what you're good at. Well, then, you'll have found yourself. You'd be living your dream."

Anthony's eyes widened as he listened to Paul talk. The advice felt like something he could really latch on to and make his own, but there was a bigger problem. "What if I'm good at several things and still figuring out what I'm passionate about?" he asked.

"Ah, I think I see the problem." Paul sighed. "Let me give you an example of how to match up passion with what you're good at. If you're passionate about fitness and teaching, you might be a personal trainer or a fitness instructor. If you're passionate about photography, you might be a wedding photographer or a fashion photojournalist. If you're passionate about business and technology, you might be an entrepreneur and build your own tech start-up company."

Anthony was silent for a moment, deep in thought. He felt inspired by Paul's words, and he knew he would have to start considering his passions and how to match them up with what he was good at. Some of his friends already had their lives and careers mapped out. They knew exactly what they wanted and the path to follow to get there.

"How do I get started?" he asked. "It seems like a lot of work, but it also sounds really exciting."

"The best way to start is to just explore. Do some research and talk to people who can help you, like your teachers. Or friends and family. But most importantly, don't be afraid to take the first step even if you're not completely sure what you're doing."

Anthony sat up straight in his chair. "Is that what you did? Took a leap of faith even though you didn't know what you were doing?" he asked, curious to hear Paul's answer.

"That's absolutely what I did. I'm good at project planning and execution. I'm good at leading and managing. But what I'm passionate about is building. Finding new and innovative project ideas and concepts. I started my own construction company, building garages and small themed hotels. My businesses have always done extremely well because I figured out my skills and matched them with my passions."

Anthony nodded, feeling that sense of excitement grow inside of him. "I think you're right. It does sound exciting."

"So, take those first steps," Paul urged him. "Explore your passions and see if they are really what you're good at, too. Then create a step-by-step plan for how to get there."

"That's the part that's what's overwhelming. Figuring out all the steps to get there," Anthony sighed.

With a smile, Paul stood up from his chair and held out his hand to Anthony. "I think I can help you there. If you'll let me."

Anthony took Paul's hand in his and stood up as well, glancing out the window at the sun setting in the distance. "I think I'd like that very much."

"Good. Then let's get to work," Paul said with an excited grin on his face.

* * * *

Mr. Omen started his life's adventure in September 1937. After high school he decided to open his first Step and Sidewalk paving business. He then joined the army reserves and served in the Engineering Battalion. In 1960 he married his wife and started a family.

In 1977 he began building the Gold Pyramid and moved

into it in 1982. He then started testing theories that the Pyramid created its own energy. Then he got involved with the Ancient Astronaut Society (known as Ancient Aliens). He became one of the leading people towards pyramid power.

He then started his own garage business with his brothers and created the slogan *NOBODY BUT NOBODY BEATS AN OMEN DEAL*. He then built several hotels, a restaurant that served healthy spaghetti and a glow-in-the-dark mini golf course.

He had visions of building the world's largest water park and war memorial. He was going to sell the pyramid to finance it but unfortunately these never broke ground.

He even went before the United Nations to get support for his largest project, Whole World Trade Mall, but it also never broke ground. Jim devoted his life to the service and happiness of others.

Chapter Nine

nthony looked up at Paul with a mixture of
excitement and nervousness as the older man stood
in front of him, a look of determination on his face.

"So, you want to know more about my life as a
businessman and entrepreneur?" Paul asked, smiling at
Anthony.

"Yes," Anthony replied eagerly. "I'm really interested in
hearing more about how you got to where you are today."

Paul nodded. "Let's take a walk. I have business
associates coming over for lunch today, and I want you to
meet them. You might get an idea of what you want to do for
your own career."

The two of them set off down the street, with Paul
regaling Anthony with stories from his life. He talked about
all the amazing business ventures he had been a part of, and
how he had started out building a construction empire before
he built the pyramid house.

"When I was young, just like you," Paul said, "I knew
that I wanted to be successful, but I didn't exactly know what

that looked like or how to get there. So, I spent a lot of time exploring my interests and figuring out what I was really passionate about."

"What is it you're really passionate about?" Anthony asked curiously.

"Well, I always loved building and creating things," Paul replied. "I loved taking an idea and bringing it to life. I was always thinking about the next big project on the horizon, and I still do. And that's what I did with my businesses—I came up with all kinds of unique projects and built them from the ground up. Everything building garage doors to installing house roofs and everything in-between."

"You've had a lifetime of hard work," Anthony replied. "But it certainly seems like all that effort has paid off."

"It really has," Paul agreed with a smile. "You've got to believe in yourself. Believe in your dreams and your ability to make them come true. Because if you believe strongly enough in what you're doing, then you'll make it happen." Paul winked at Anthony and tapped his temple with one finger. "That's what makes our country so great. The American dream—it's not just a myth. It's real, if you make it so. And that starts with having the right attitude and mindset."

As Anthony listened to Paul speak, he began to feel inspired. He knew that he wanted something special in his life—like Paul had with the pyramid house and his businesses and family-and his new mentor made it easier to believe that if he put in the work and followed his own passions, then anything was possible.

Paul had helped so many people in—he had only met a few of them, and he knew there were more out there—and Anthony knew he wanted to help other people, too. The question was: what the best path was to take to build his own businesses while helping others.

Anthony thought about his dad and mom. They were hard-working people, but they had a family to support and were always focused on doing well at their jobs and building a strong future for their kids. They wanted to see him succeed, of course, but they always encouraged him to be a doctor or a lawyer or something else that was high-status and secure.

He wondered whether his parents had ever had big dreams for themselves when they were young. Did they sacrifice their personal dreams because the needed to provide well for him and his sister? What if he desired a more unconventional career path?

Anthony didn't want a safe, ordinary life. A life where he followed someone else's idea of success wasn't enough for him. He wanted to be passionate about his work, and he wanted to find his own dream and not be frightened to go all out just because his ideas might be unconventional or challenging.

When he looked at Paul Omen, he saw an adventurous man. Someone who had taken risks and followed his own dreams. He knew that he wanted to do something similar, and, after meeting Paul, Anthony realized others had blazed their own trails to their dreams instead of taking the safe, conventional route.

If they did it—if Paul did it—then maybe he could, too.

They walked outside to a large patio area outside the pyramid house. There were so may visitors sitting at picnic tables around a concession stand, chatting and enjoying their time in the sunshine. A tall man, seated at a picnic table with two other men and two women, spotted Jim and waved him over.

"Jim," The man said, as he stood up. He was a few years older than Jim, with dark hair and a stockier build. He looked like he had worked hard all his life, a real hands-on kind of guy. He extended a hand to Jim that was calloused from hard work. "Hope you don't mind. We couldn't resist the hot dogs, so we started without you."

"Don't mind at all, Steve," Jim replied with a wide smile, as he clasped his friend's hand. "How have you been?"

"Great. We've been looking forward to this lunch all week." Steve gestured to the other people at the table, then he smiled at Anthony. "And I see you've brought a friend, too. I'm Steve, by the way, and this is my wife, Lisa," he said, pointing to a woman with long brown hair who was sipping on a cold drink. "We've known Jim for ages. We're part of his Gold Pyramid Mineral Water distribution network."

Anthony extended a hand to shake, and Steve gave his hand a firm, friendly squeeze. "Nice to meet you."

Jim gestured for Anthony to sit as he pulled up a chair so he could sit at the head of the picnic table. "Everyone this is Anthony. I met this enterprising, curious young man when his family visited our Gold Pyramid one day, and he just kind of stuck."

Lisa laughed and smiled at both Anthony and Jim. "I have a feeling you stuck around because of Jim. He's like a planet with enormous gravitational pull. You can't help but be drawn to him."

Jim waved the compliment away with a self-deprecating smile and a dismissive wave. "Now, Lisa, you're making me blush."

"That would be a first, my friend," Lisa said with a giggle. "Everyone knows you're a man who makes things happen. You're the one who convinced our network partners that Gold Pyramid was worth working with, and that we just needed to expand your distribution."

Jim chuckled. "And I'm so glad you did. Young Anthony has really taken to meditating here at the Gold Pyramid. He used to suffer horrible migraines but after talking with a few of our friends, who meditate in pyramids, he decided to try it himself to see if it would work.

"And?" Steve asked, eager to know more. "How is it going?"

Anthony shrugged. "It's going great. I've been meditating for about a few months now, and the migraines are just about gone completely. I can't believe how much better I feel."

"That's fantastic to hear," Steve replied.

"Yeah, and that's not all. Mr. Omen—I mean, Paul—has been helping me a lot. I'm going to college in the fall, and he's been sort of mentoring me. Helping me figure out what kind of career I might want to prepare for. He's given me so many ideas and tools, and he's just a wealth of knowledge," Anthony continued.

Melissa, who had kind, smiling eyes, spoke up. "I'm glad to hear you're getting some help from Paul. He's a great guy. He's helped a lot of people with business ideas."

Steve nodded. "Yeah, he really is the best at what he does and everyone in our network would agree. We've all been very lucky to be able to work with him."

Lisa gazed at Anthony, and her voice filled with warmth. "Maybe we can help, too. We're pretty good listeners and I'm sure we all remember what it's like to be young and figuring out what we want to do in life. Why don't you eat with us and tell us about what you have in mind? We're all ears."

"That's exactly what I was hoping you'd say," Paul replied, smiling at the group. "Anthony has a lot of options, pretty much the whole world is open to him, if he wants it."

Anthony smiled, touched by the offer. "That would be great. Paul was telling me how important it is to find what you're good at and what you're passionate about, and then follow that path. It seems like a great idea. I just starting to figure out what things I like to do and what motivates me to get up in the morning, and I'm thinking more about how that can translate that into helping other people."

Melissa turned and gave Jim an admiring look. "Sounds like you're the right person to mentor him, Jim. He can benefit from your wisdom and experience." She turned back to Anthony. "Our friend here has always had big, big dreams. You know, when Jim was building the Gold Pyramid, a lot of people laughed at him and his ideas. But look at him, now. The Pyramid House is a big tourist attraction, and he gives speeches to groups of people about it all the time."

"That's right," Lisa agreed. "He did it. Not only did he build the Gold Pyramid, but he has also built motels, and all kinds of different businesses."

Paul shrugged off the compliments again with a humble smile. "Success is what you make of it. And it's time my wife and I gave back to the country we love so much. We're trying to figure out how to do that but, in the meantime." He clapped Anthony on the shoulder. "It's a real privilege to be working with this bright young man."

Steve beamed at the group, clearly proud of his friend. "So, let's help Anthony, here, start brainstorming his future. What do you think, guys?"

"Sounds like a plan to me," Melissa agreed. "So, Anthony,

what kinds of things do you think you're good at?"

Anthony felt his heart pounding with excitement at the prospect of having such a supportive group to help him on his journey. "I can't thank you enough," he said. "What am I good at? Well, my friends say I'm good at talking to people. I can just kind of put them at ease and help them feel better about themselves and the situation they're in. I guess it's partly because people open up to me so easily."

Lisa nodded thoughtfully. "That makes sense, I think anyone who is good at helping others would have a natural ability to do that."

Anthony accepted a couple of hot dogs that one of the servers had brought out, and he took a bite. The food was delicious, and he started to relax, a bit, as the conversation carried on around him.

"So, what do you like to do when you're not eating hot dogs?" Melissa asked, with a smile. "If you tell us what you like to do, and what you think you're good at, that would be a good place to start."

"I love to read," Anthony replied, thinking about all the books he had devoured over the years. "I also love to workout. I played lots of sports in school—baseball, basketball, and soccer. Staying fit is really important to me."

Steve took a small notepad from his jacket pocket and began writing things down. "Ok, loves to read. Health-conscious, fit, loves sports and likes to help people." He looked up at Jim and Anthony. "Is it just me or do you all see the potential for a career in health and wellness taking shape?"

Jim pursed his lips, thinking. "For sure. There's also his personal experiences of meditating in pyramid structures. He has first-hand knowledge that the pyramid can help people feel grounded and centered. Not to mention that it helped his migraines. I'd say that knowledge is definitely useful."

"Absolutely!" Paul agreed. "And he has a very balanced point of view—he's not obsessed with taking care of only one aspect of his life or another, but many aspects. And from what I've seen, so far, he wants to help people. So, maybe look into studying to be a fitness trainer or something like that."

"I would like to start my own business of some kind," Anthony continued. "Something in the wellness industry sounds pretty awesome. I could help people with their fitness or address their physical challenges."

"Maybe be a physical therapist," Lisa suggested. "Or a chiropractor or an osteopath. There are so many things you could do, Anthony."

"I know it sounds like a big decision to make," Jim said, "but honestly, if you're passionate about helping people, and you have several areas of interest that all tie in together, there's no reason why you can't take your time to choose exactly what it is that's right for you. You could also do some research and see what kind of jobs exist in the industry you're interested in. You could also start by looking into earning a college degree, or maybe some qualifications that would get your foot in the door, then you can figure out which direction to go from there."

Anthony nodded slowly, thinking about everything his new mentor said. "I really appreciate all the support you're giving me. I have so much to think about. I go off to college in a few weeks, and I feel so much pressure because a lot of my friends know exactly what they want to do already."

"Well, now you have a great starting point," Steve said. "Keep searching for your strengths. You already know that you're interested in helping people. So go out there and see what's available! You might even be able to find some areas where it looks like the opportunities are."

"We're forgetting two really important pieces of the puzzle," Lisa said. Everyone looked at her, expectantly. "First, stay true to yourself. Second, keep Jim Omen as your mentor. You can always count on him to point you in the right direction, and help you achieve your goals faster."

Jim smiled at Anthony and gave him a fatherly pat on the back. "And it will be my pleasure and privilege to help you get there, Anthony. Whatever you choose to do with your life, I promise that I'll always be here to support you. No matter what."

Chapter Ten

Four Years Later

nthony gazed out the window as his plane taxied into Chicago-O'Hare airport. As he exited the plane and made his way through the busy airport, Anthony felt a rush of excitement and anticipation. As he looked around at all the people rushing to make their flights—business people traveling in suits, couples and families heading off on vacation, backpackers and college students about to embark on a new adventure—Anthony knew that something had changed in him.

He made his way to the baggage claim area, eyes scanning the crowds for his parents and younger sister, who had come to pick him up. As they finally spotted each other, Anthony felt a wave of love and pride wash over him.

"It's so great to be home!" he said, giving his family tight hugs.

His parents smiled at him proudly, and his sister beamed up at him with an eager grin on her face. "Did you bring me some presents?" she asked.

Anthony laughed. "Of course, I did! But you'll have to wait until we get home to find out what they are."

His mother kissed him on the cheek and then hugged him tightly. "Oh, sweetie, we're so glad you're home. We can't wait to hear more about your summer in Europe. The pictures you sent were lovely."

"I know you had a great time over the summer," his dad said, peering at him with a look of quiet pride and admiration. "And I'll bet you're itching to get your career started."

Anthony picked up his backpack, filled with dirty laundry and souvenirs. "Yeah, I'm excited to get busy. I met a lot of people at that Wellness conference and trade show in Frankfurt. I made loads of good connections."

"Well, we're so proud of you, son," his dad said. "I know it has been a lot of work for you, but if you keep working hard, it's all going to pay off."

"Yes, sir. And I'm going to do everything I can to make you all proud of me," Anthony replied.

As they climbed into the car and drove home with his family, Anthony felt an overwhelming sense of gratitude for all that he had been given. He was filled with a sense of purpose, and he knew that with the support of his family and friends, he would be able to achieve his dreams.

But there was one person he knew he had to see as soon as possible: Paul Omen. He needed to seek out Paul's wisdom and guidance, and he couldn't wait to see his long-time mentor in person and tell him all about his summer in Europe traveling and working with other people who were committed to improving people's health and wellness.

After they arrived at home, Anthony sat down with his family for a lunch of his favorite comfort foods: homemade pizza and fresh salad. As they ate, Anthony told them all about his adventures over the summer, regaling them with stories of the fitness conventions he'd attended in Frankfurt and London, as well as all the amazing new connections he'd made with other health professionals.

Jenna perked up when Anthony brought out the souvenirs, he had brought home for her: handmade jewelry from the small villages he'd visited in Italy, a beautiful pastel watercolor of the canals in Venice that he had bought from an artist on the street, and even a little figurine of a tulip that reminded him of Amsterdam.

"Oh wow!" Jenna said excitedly. "You really got me something special." She hugged Anthony tightly. "Thank you so much!"

"I'm glad you like it," he replied, smiling at her fondly. "The summer was amazing, and I want to make sure that you know that I appreciate all the love and support you've given me over the years." He leaned over and gave Jenna a kiss on the cheek.

For his mom, he had brought home souvenirs from the English countryside: packs of scented soaps from Bath, a tea towel with an adorable print of a village green, and even a small pottery cup that had been hand-painted with delicate pink flowers and thistles.

"Oh, I love these!" his mom said excitedly as she sifted through the different souvenirs. "Thank you, honey.

His dad got a bottle of whiskey from Scotland, along with some beer steins from Germany.

"Thank you so much, son," his dad said, pulling Anthony into a tight hug. "These are wonderful gifts."

Anthony smiled. "They're nothing compared to the gifts of love and support and encouragement that you've given me over the years. I couldn't have gotten to this point without you."

"It's our pleasure, sweetie," Mom beamed at him. "We're just so proud of you."

"Are you free to go play a round of golf this afternoon? Just me and you?" Dad asked. "I know a few people at the club who would be excited to see you."

Anthony shook his head. "Not today. Tomorrow, maybe. There's someone I need to visit this afternoon. I promised I would stop by as soon as I got home."

"Oh, who is that?" his mom asked. "Anyone we know?"

"Yep," Anthony replied, his voice full of excitement. "Paul Omen. He has been an awesome mentor and coach. I've learned so much from him about how to become an entrepreneur and how to achieve my goals. He's one of the reasons I've come so far—and I want to share with him all that has happened this summer."

"I think it's great that you have a business mentor who believes in you and helps you stay focused on what's important," Mom said, patting his hand. "You should definitely spend some time with him today. I heard the Pyramid had a fire, but Mr. Omen is going to rebuild the house to host events there."

Anthony glanced at the clock on the stove. "I can't wait to talk to him. I've got to get going. I'll see you all later." He gave them all one last hug before heading out the door.

As he drove to Wadsworth, back to the Gold Pyramid house where his career journey first started, he watched the suburban landscape change from neatly manicured lawns and basketball hoops in the driveways to more scenic and pastoral areas that were the hallmarks of rural Illinois. He thought about how much he had changed since he first visited at the pyramid house and reflected on all the amazing experiences that had led him to this point.

It was odd to think that he almost missed such a life-changing day altogether. He hadn't wanted to go on that outing to the Gold Pyramid with his family. He had been tired and grumpy, and he just wanted to stay home. But somehow, his parents' insistence that they spend time together during the day trip had pushed him out of his funk, and that set in motion a journey that would change his life forever.

Yolanda Fierro

As he pulled up to the Gold Pyramid house, Anthony felt a rush of excitement at seeing Paul again after so long. He stepped out of his car and walked up the long driveway, smiling as he remembered all of Paul's stories and lessons. He was eager to tell him about his own success and see what he had in store for the future.

Paul greeted Anthony with a wide smile and a warm hug. "And now comes home the traveling soul who has seen amazing sights and embarked on a journey to greatness," he said, gesturing dramatically. "Welcome home, Anthony."

Anthony laughed and shook his head in response. "It's really good to see you, Paul—I mean that sincerely. It's been a while since we've talked."

"That it has, my boy. Come on in. Let me get you something to drink. We've had a bit of a heat wave, lately, but that's late summer in Illinois, isn't it?

"Sure is. Maybe we could score some Gold Pyramid Mineral Water? I think I know a guy who could hook us up with some of that primo elixir." Anthony's eyes twinkled with mirth.

"Ha! Yes, of course. I think we can manage to do that," Paul said with a laugh, motioning towards the back of his house. "In fact, I've got some in the fridge right now. But I also have some beer if you'd like. Now that you're of drinking age."

"I wouldn't say no to a beer. I've got some good news to

share with you." Anthony followed Paul into the kitchen and sat down at the table.

"Good news, huh? What's up?" Paul asked, setting a bottle of premium ale down in front of Anthony.

"Yeah, but first, I want to hear how you've been. Any new adventures, insights, or business deals to share with me?"

"Oh, plenty of adventure! We have some new Egyptian treasures in the house, as well as some new artifacts for King Tut's tomb."

Anthony's eyes lit up. "Oh, you have to show me! I'm so excited to see what you've been working on."

"Well, you know how I love to share new treasures with people," Paul said, a twinkle in his eye. "In fact, you'll be the first to see a few of the new treasures." He led Anthony to the main Egyptian display room.

They stopped at a large golden chest and Paul lifted the lid, revealing gleaming cabochons of lapis lazuli. Anthony gasped in delight as his eyes were drawn to one particularly beautiful piece.

"Wow, these are gorgeous," he said, turning to Paul with a smile. "You're always coming up with something amazing."

"We have a new replica of Tutankhamun's gold chest. Notice the engravings on the outside. They're exact replicas of the ones on the original. And the king's cartouche is engraved in gold on the lid. It's quite extraordinary."

Anthony nodded, hypnotized by the beauty of the piece.

"You're totally right—it's amazing! This really is a treasure trove."

Paul smiled. "I'm glad you think so, my boy. But there's more—come over here to this display of statues."

Anthony followed Paul to the statue display and gasped again as he saw the beautiful figurines carved from lapis lazuli and other stones. Then, a bust sporting an elaborate gold collar caught his eye. "This is exquisite. What is this?"

"That is an Anubis collar, a replica of one that belonged to Nesi-Ameni, a scribe and royal chancellor for the Egyptian king Tutankhamun. It is carved from lapis lazuli and gold."

"That's incredible," Anthony shook his head in wonder. "I love how you go to every length to make sure these pieces are as authentic and true as possible. It's amazing how detailed they are."

Paul grinned proudly. "Well, we do what we can for authenticity, but also with our own flair and personality," he said, sweeping his arm towards the rest of the statues. "I think it's important to put your own mark on your work."

"I completely agree," Anthony said, nodding. "It's what makes a collection unique and special. And it really takes passion to pull off something like this."

Paul laughed heartily. "There is no doubt about that! Now, tell me how you got on in Europe. I want to hear more about those conferences you attended."

"The conferences were amazing. I met some people

who are planning a new kind of health, wellness, and fitness experience for people. They're creating a one-stop shop of physical, mental, emotional, and spiritual wellness under one roof."

"That sounds intriguing. What kind of experience would it be?"

"Well, they're looking at a model that's very similar to what traditional healers have been doing for centuries—but with a modern twist. They want to harness the power of ancient healing techniques and techniques from the non-Western world, like acupuncture, yoga, and reiki, and combine them with Western techniques like physical fitness, mental wellness, and meditation. They want to create a complete mind-body experience that caters to all aspects of the individual."

"That sounds fascinating," Paul said, rubbing his chin in thought. "It's definitely something I'd like to learn more about. And it sounds like you're really interested in this new wellness model too."

"I am. When I told them about the power of pyramids, and my experiences here at the Gold Pyramid house, they were really interested in learning more about it. They were especially excited when I told them how meditating in a pyramid structure helped my migraines."

Paul raised an eyebrow in curiosity. "Well, I'm glad to hear you're still finding success with pyramid power. Sounds like this new wellness model could be a great place for you to put your knowledge and expertise to use."

"I agree," Anthony said excitedly. "They invited me to join them. They're looking for people to work as fitness trainers and yoga instructors. My degree in kinesiology is exactly what they're looking for, and they think the power of pyramids could be a game changer for their business."

"That sounds like an incredible opportunity," Paul said, getting a little misty-eyed "I can't wait to see what you do with it. I knew from the moment we met that you were someone special. I'm so proud of you and all that you've accomplished."

Touched by Paul's words, Anthony beamed. "Thank you, Paul. I couldn't have done it without your support and encouragement. I'm glad you saw something in me that I didn't even see in myself. I can't thank you enough for all of your guidance and support. You're a remarkable man, and I'm so lucky to call you a mentor and a friend."

Paul smiled. "I think you're the one who is amazing," he said, placing his hand on Anthony's shoulder. "You've come so far in such a short time, and I know that there are great things ahead for you. I'm honored to be able to share your journey with you.

"Thank you, Paul," Anthony said, tearing up. "I couldn't have done it without you."

* * * *

The two friends embraced, and for a moment the world stood still. In that moment, they knew that their bond would last long after their time in the Gold Pyramid house. They had both found something within each other that had been missing in their lives, and that would never change. It was a truly magical moment, one that they would treasure forever. And they could only hope that the world would learn to value pyramid power as much as they did.

As Anthony looked back on all that he had accomplished, he felt a sense of pride and accomplishment. He knew that his journey was just beginning, and there were many exciting opportunities ahead for him. But most importantly, he knew that none of it would have been possible without the support and guidance of Paul Omen, his mentor and friend.

With the knowledge and wisdom that Paul had shared with him, Anthony was now empowered to share his own story of pyramid power with the world. He knew that there were many people out there who sought healing and wellness, and he felt an overwhelming sense of purpose in bringing the power of pyramids to them.

Anthony was so inspired by Mr. Omen that he now hosts the Global Pyramid Conference where people come from around the world to attend and share their stories and wisdom about the Pyramids.

With unwavering faith and an unshakable belief in himself, Anthony knew that he was destined for great things.

And he was determined to use the power of pyramids to make his dreams a reality.

Epilogue

It's been seven years since Jimmy stepped up to carry on James Onan, Sr's vision to make the pyramid a place for people to come get a small glimpse of Egypt and feel the healing energy that the pyramid contains.

We continued on, even through a devastating fire in 2018 that was caused by a worker removing the gold like coating from the exterior of the building. He struck a nail and it caused a spark that started a fire in the insulation on the inside of the wall. The damage to the exterior was significant, but the inside was devastated by the water from the moat surrounding the house that firefighters used to put the fire out.

In trying to create revenue to help maintain the property while fire restorations were underway, The Gold Pyramid Mineral Water and Vodka brands were launched. Having the water tested it was found to have high levels of calcium, magnesium and potassium, a natural electrolyte that flows up from under the middle of the house.

A 4,000 sq. ft. tent was erected outside to accommodate large events such as weddings, concerts, car shows and fundraisers.

Our goal is to raise money through the sales of the water, vodka and events to help build the Brain Medicine Institute & Educational-Research Center in Waukegan, Illinois on several acres of Jimmy's personal property that he has donated to this project in honor of his dad.

This center will provide excellent treatment and studies for all types of brain disorders. Dr. Ricardo Senno an expert in brain disorders has taken on the challenge to help this project become a reality.

If you would like more information or make a donation to this cause you can go to their website: Brainmedicineinstitute.org or contact Dr. Richard Senno at senno@brainmedicineinstitute.org. to find a cure for the deadly diseases Alzheimer's and Dementia that affects millions of people and took James Onan Sr's life in July of 2023.

He would say, "The possibilities here are endless, and someday everyone will know the Power of the Pyramid House."

About Yolanda Fierro

My father was in the military and was transferred to Illinois where he met my mother. After they married, he was transferred again to San Diego, California where I was born. Then moving back to Lake County Illinois where they would have my younger sister and call home.

Unfortunately, what was to be a happy ever-after tale would turn out to leave me homeless at the age of thirteen.

I finally moved in with my grandparents and began my journey in adulthood.

I married young and gave birth to my son by the age of twenty-one then divorced and remarried having my beautiful daughter at thirty-one. But that relationship didn't last so I started on a new road to see what God wanted me to do.

Being a single mother of two, I pushed myself to continue learning new ways to support my family and obtained a Cosmetology License and started working at a local beauty salon where I was a makeup artist and represented our salon in providing services to local Chicago sports team photo shoots.

Yolanda Fierro

Not making enough the support my family, in 2005 I took on another job learning to lease apartments and obtained a Leasing Agent license. In 2010, I became a Licensed Real Estate Salesperson, and in 2011 a Real Estate Broker then in 2011 a Licensed Real Estate Managing Broker.

Still looking for my path, in 2013 I became a Licensed Health and Life Insurance agent. Even with all these accomplishments I still felt like my true path was still out there waiting for me to find it.

Then Jimmy reached out to me requesting my help in a new company. As it turned out, it was much more than that. Knowing the family for over thirty years, listening to the stories and personally working with Jim Onan, Sr. and Jim Onan, Jr. I've seen and experienced firsthand the wonders of the Pyramid house and the effect it has on people—how it changed their lives including my own.

When I began my journey with the Onans, it was to assist in creating the Pyramid Water brand, but it evolved into a whole new path. Learning the mysteries of the Gold Pyramid opened me up to new way of life. The positive energy I felt in the project was amazing and so were all the wonderful people I met along the way.

Now as executive director for Gold Pyramid Brands, I am instrumental in the day-to-day operations of the Pyramid house events and tours. I even get to dress up as Cleopatra. I oversee the Pyramid Bottling, Pyramid Vodka, Onan Storage, Onan Senior Housing and commercial buildings. I also have

Club Tiki and Isabella's as part of the Jimmy Onan enterprise.

We never know where God will take us, but if we open our hearts to Him, the possibilities are endless. My experiences at the Gold Pyramid inspired me to tell the story of the *Passion of the Gold Pyramid* in the hopes that you will make the journey to experience it and feel the unique, life-changing energy.

Visit www.goldpyramid.com